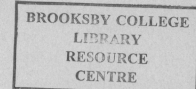

20/1

simple methods for aquaculture

WITHDRAWN

POND CONSTRUCTION
for freshwater fish culture

building earthen ponds

Text: A.G. Coche
J.F. Muir
T. Laughlin
Illustrations, book design
and layout: T. Laughlin
Artwork: E. D'Antoni

FOOD AND AGRICULTURE ORGANIZATION OF THE UNITED NATIONS
Rome 1995

David Lubin Memorial Library Cataloguing-in-Publication Data

FAO, Rome (Italy)

Pond construction for freshwater fish culture:
   building earthen ponds
(FAO Training Series, No. 20/1)
ISBN 92-5-102645-9

1. Fish ponds   2. Fish culture
I. Title   II. Series

FAO code: 44   AGRIS: NO1 M12

P-44

ISBN 92-5-102645-9

THE AQUACULTURE TRAINING MANUALS

The training manuals on simple methods for aquaculture published in the FAO Training Series are prepared by the Inland Water Resources and Aquaculture Service of the Fishery Resources and Environment Division, Fisheries Department. They are written in simple language and present methods and equipment useful not only for those responsible for field projects and aquaculture extension in developing countries but also for use in aquaculture training centres.

They concentrate on most aspects of semi-intensive fish culture in fresh waters, from selection of the site and building of the fish farm to the raising and final harvesting of the fish.

FAO would like to have readers' reactions to these manuals. Comments, criticism and opinions, as well as contributions, will help to improve future editions. Please send them to the Senior Fishery Resources Officer (Aquaculture/Publications), FAO/FIRI, Viale delle Terme di Caracalla, 00100 Rome, Italy.

The following manuals on simple methods of aquaculture have been published in the FAO Training Series:

Volume 4 — Water for freshwater fish culture
Volume 6 — Soil and freshwater fish culture
Volume 16/1 — Topography for freshwater fish culture: Topographical tools
Volume 16/2 — Topography for freshwater fish culture: Topographical surveys
Volume 20/1 — Pond construction for freshwater fish culture: Building earthen ponds
Volume 20/2 — Pond construction for freshwater fish culture: Pond-farm structures and layouts

Two final volumes are being prepared:

Volume 21/1 — Management for freshwater fish culture: Ponds and water practices
Volume 21/2 — Management for freshwater fish culture: Farms and fish stocks

HOW TO USE THIS MANUAL

The material in the two volumes of this manual is presented in sequence, beginning with basic definitions. The reader is then led step by step from the easiest instructions and most basic materials to the more difficult and finally the complex.

The most basic information is presented on white pages, while the more difficult material, which may not be of interest to all readers, is presented on pages with a grey or light blue background.

Some of the more technical words are marked with an asterisk (*) and are defined in the Glossary on page 353.

For more advanced readers who wish to know more about fish-farm construction, a list of specialized books for further reading is suggested on page 355.

CONTENTS

CONTENTS, continued

CONTENTS, continued

CONTENTS, continued

CONTENTS, continued

CONTENTS, continued

CONTENTS, continued

TABLES AND GRAPHS

1 GENERAL BACKGROUND

10 Introduction

1. A large part of the world's fish culture production relies on the use of freshwater ponds which hold and exchange water, receive fertilizer or feed, and allow for holding, rearing and harvesting of fish. The proper preparation and construction of such ponds and their associated structures are essential for successful fish farming. Good ponds should be inexpensive to construct, easy to maintain and efficient in allowing good water and fish management.

2. The purpose of this manual on *Simple Methods for Aquaculture* (**Pond construction for freshwater fish culture**, *FAO Training Series*, Volumes **20/1** and **20/2**) is to provide the basic knowledge needed to build good, efficient and reliable pond systems. Both volumes of this manual should ideally be used together with earlier manuals on *Simple Methods for Aquaculture* (**Water for freshwater fish culture**, *FAO Training Series*, **4**; **Soil and freshwater fish culture**, *FAO Training Series*, **6**; **Topography for freshwater fish culture**, *FAO Training Series*, **16/1** and **16/2**). The next manual in this series will deal with pond and fish management (**Management for freshwater fish culture**, *FAO Training Series*, **21/1** and **21/2**).

11 Features of a fish pond

1. Although there are many kinds of fish ponds, the following are the main features and structures associated with them in general:

- **pond walls or dikes**, which hold in the water;
- **pipes or channels**, which carry water into or away from the ponds;
- **water controls**, which control the level of water, the flow of water through the pond, or both;
- **tracks and roadways** along the pond wall, for access to the pond;
- **harvesting facilities** and other equipment for the management of water and fish.

Note: in this manual, a fish pond is defined as an artificial structure used for the farming of fish. It is filled with fresh water, is fairly shallow and is usually non-flowing. Tidal ponds, reservoirs, storage tanks, raceways and fish farm tanks are not included.

Parts of a fish pond

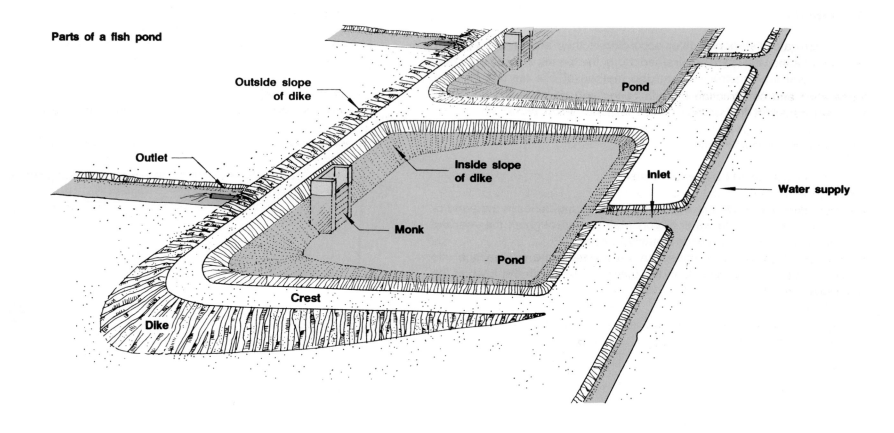

Outside slope of dike

Outlet

Inside slope of dike

Inlet

Water supply

Monk

Pond

Pond

Crest

Dike

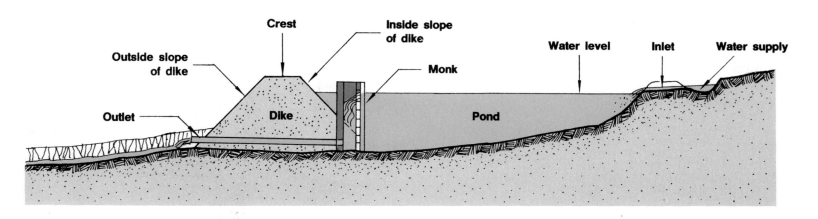

Crest

Inside slope of dike

Water level

Inlet

Water supply

Outside slope of dike

Monk

Outlet

Dike

Pond

3

12 Different kinds of pond

1.   Freshwater fish ponds differ according to their source of water, the way in which water can be drained from the pond, the material and method used for construction and the method of use for fish farming. Their characteristics are usually defined by the features of the landscape in which they are built. Ponds can be described as follows.

According to the water source

2.   Ponds can be fed by **groundwater**:

(a) **Spring-water ponds** are supplied from a spring either in the pond or very close to it. The water supply may vary throughout the year but the quality of the water is usually constant.

(b) **Seepage ponds** are supplied from the water-table by seepage into the pond. The water level in the pond will vary with the level of the water-table.

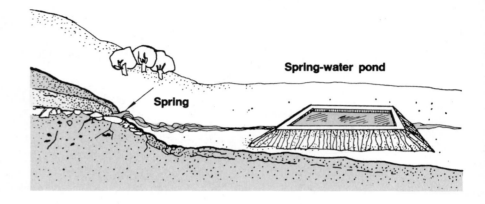

Spring-water pond

Spring

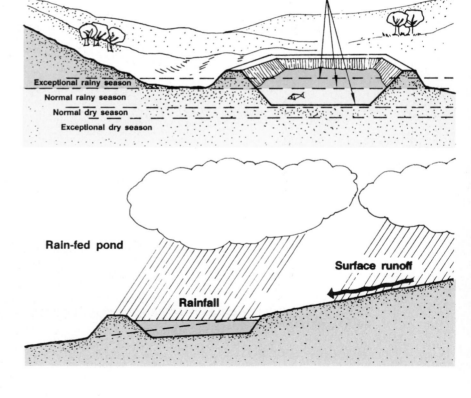

Seepage pond
(water table varies)

Exceptional rainy season

Normal rainy season

Normal dry season

Exceptional dry season

3.   **Rain-fed ponds** are supplied from rainfall and surface runoff. No water is supplied during the dry season. These ponds are often small depressions in impermeable soil, with a dike built at the lower side to retain more water.

Rain-fed pond

Surface runoff

Rainfall

4.   **Ponds can be fed from a water body** such as a stream, a lake, a reservoir or an irrigation canal. These may be **fed directly** (e.g. **barrage ponds**), by water running straight out from the water body to the ponds, or **indirectly** (e.g. **diversion ponds**), by water entering a channel from which controlled amounts can be fed to the ponds.

Ponds fed from a water body

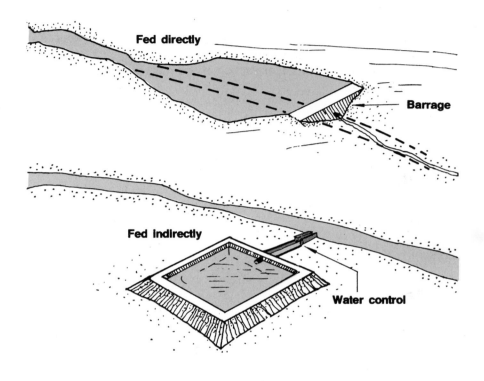

Fed directly

Barrage

Fed indirectly

Water control

5.   **Pump-fed ponds** are normally higher than the water level and can be supplied from a well, spring, lake, reservoir or irrigation canal, by pumping.

Pump-fed pond

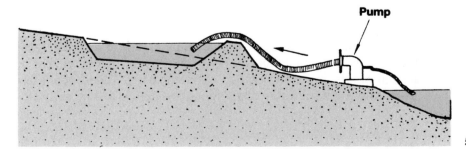

Pump

According to the means of drainage

6. **Undrainable ponds** cannot be drained by **gravity***. They are generally fed by **groundwater** and/or **surface runoff**, and their water level may vary seasonally. Such ponds have two main origins.

(a) They may be dug in swampy areas where there is no source of water other than groundwater.

(b) They may result from the extraction of soil materials such as gravel, sand or clay.

Undrainable ponds

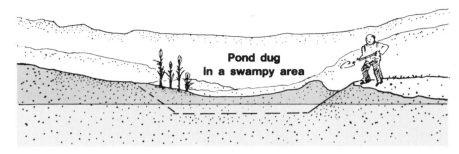

Pond dug in a swampy area

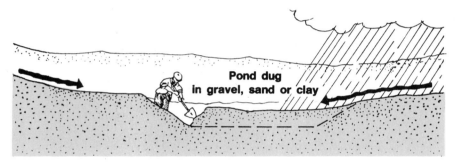

Pond dug in gravel, sand or clay

7. **Drainable ponds** are set higher than the level to which the water is drained and can easily be drained by **gravity***. They are generally fed by surface water such as runoff*, a spring or stream, or are pump-fed.

Drainable pond

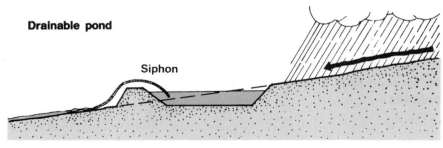

Siphon

8. **Pump-drained ponds** may be drainable by gravity to a certain level, and then the water has to be pumped out. Other ponds, similar to undrainable ponds, must be pumped out completely. These ponds are only used where groundwater does not seep back in to any extent.

Pump-drained pond

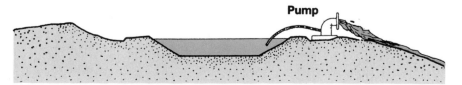

Pump

According to the construction materials

9.   **Earthen ponds** are entirely constructed from soil materials. They are the most common, and you will learn primarily about these ponds in this manual.

Earthen pond

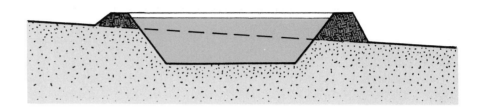

10.   **Walled ponds** are usually surrounded by blocks, brick or concrete walls. Sometimes wooden planking or corrugated metal is used.

Walled pond

Brick, block or concrete

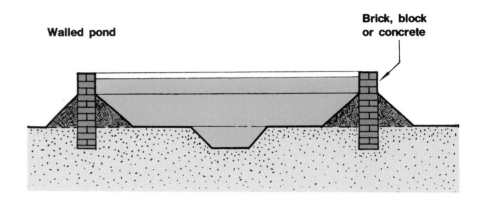

11.   **Lined ponds** are earthen ponds lined with an impervious material such as a plastic or rubber sheet.

Lined pond

Plastic or rubber sheet

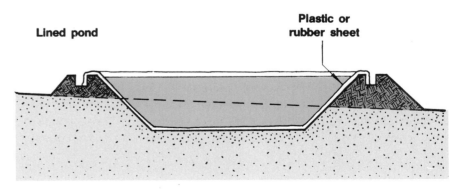

According to the construction method

12. **Dug-out ponds** are constructed by excavating soil from an area to form a hole which is then filled with water. They are usually undrainable and fed by rainfall, surface runoff* or groundwater.

Dug-out pond

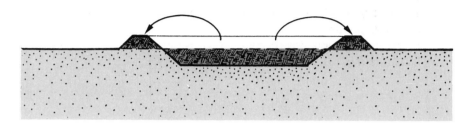

13. **Embankment ponds** are formed without excavation by building one or more dikes above ground level to impound water. They are usually drainable and fed by gravity* flow of water or by pumping.

Embankment pond

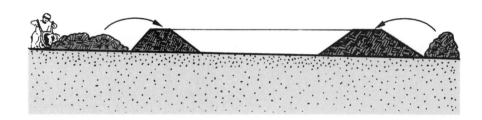

14. **Cut-and-fill ponds** are built by a mix of excavation and embankment on sloping ground. They are usually drainable, and water, which is impounded within the dikes, is fed by gravity or by pumping.

Cut-and-fill pond

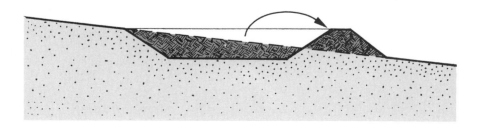

According to the use of the pond

15. There may be different types of pond on a fish farm, each used for a specific purpose:

- **spawning ponds** for the production of eggs and small fry;
- **nursery ponds** for the production of larger juveniles;
- **brood ponds** for broodstock rearing;
- **storage ponds** for holding fish temporarily, often prior to marketing;

- **fattening ponds**, for the production of food fish;
- **integrated ponds** which have crops, animals or other fish ponds around them to supply waste materials to the pond as feed or fertilizer;
- **wintering ponds** for holding fish during the cold season.

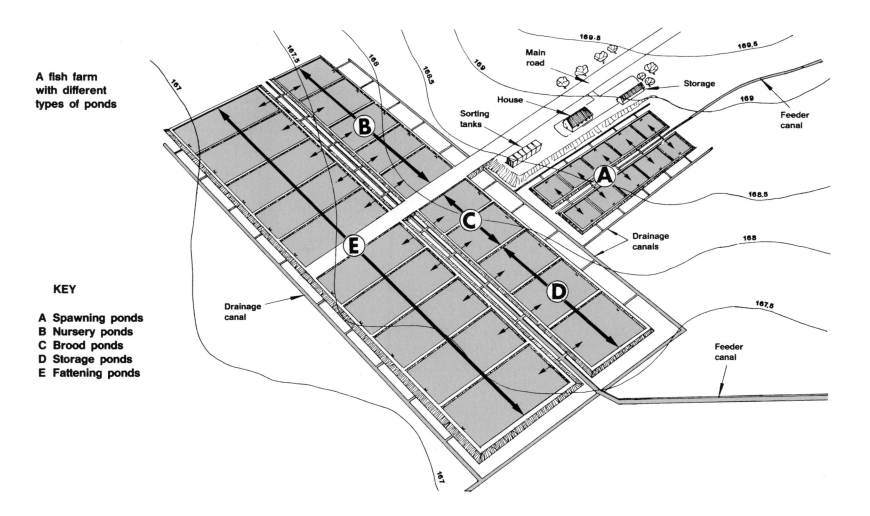

A fish farm with different types of ponds

KEY

A Spawning ponds
B Nursery ponds
C Brood ponds
D Storage ponds
E Fattening ponds

9

TABLE 1

Basic types of freshwater ponds

BASIC TYPE OF POND (subtypes, see Section 17)	MAIN WATER SUPPLY						DRAINING			CONSTRUCTION METHOD		
	Groundwater		Surface water	Water body		Pumped						
	Seepage	Spring	Rainfall and surface run-off	Direct	Indirect	Various sources	Undrainable	Drainable	Pumped	Dug-out	Embankment	Cut-and-fill
SUNKEN POND single water supply any combination of supply	●	●	●		●		●		○	●	○	○ sloping ground
BARRAGE POND without diversion canal with diversion canal in series		○	●	●	●		○	●	○		● dam	
DIVERSION POND in series in parallel			○		●	●		●	○		● flat ground	● sloping ground

● Most common
○ Less common

13 Three basic pond types

1. As you have just learned, there are many types of pond. As shown in **Table 1**, they can be conveniently grouped into three basic types depending on the way the pond fits in with the features of the local landscape.

Sunken pond

2. The pond floor is generally below the level of the surrounding land.

3. The pond is directly fed by groundwater, rainfall and/or surface runoff. It can be but is not normally supplemented by pumping.

4. The sunken pond is undrainable or only partially drainable, having been built either as a **dug-out pond** or to make use of an **existing hollow** or depression in the ground, sometimes with **additional embankments** to increase depth.

Examples of sunken ponds

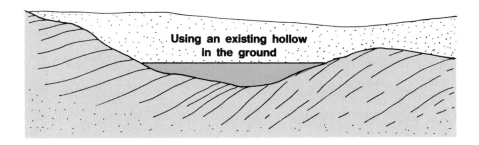

Using an existing hollow
in the ground

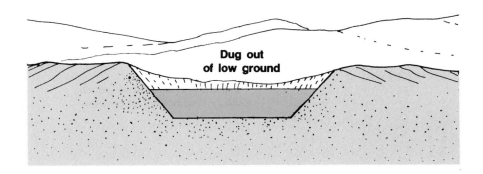

Dug out
of low ground

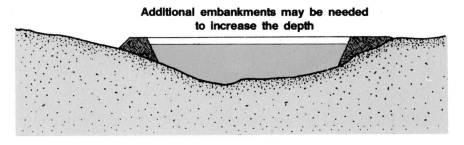

Additional embankments may be needed
to increase the depth

**Examples of sunken ponds
built on the bottom of a valley**

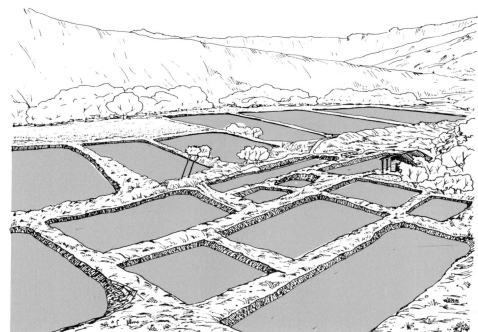

Barrage pond

5. They are created in the bottom of a valley by building a **dam** across the lower end of the valley. They may be built in a series down the valley.

6. The barrage pond is drainable through the old river bed.

7. If large floods are present, the excess water is normally diverted around one side of the pond to keep the level in the pond constant. A **diversion canal** is built for this purpose; the pond water supply is then controlled through a structure called the **water intake**.

8. Directly fed from a nearby spring, stream or reservoir, the water enters the pond at a point called the **inlet** and it flows out at a point called the **outlet**.

9. To protect the dike from floods, a **spillway** should be built.

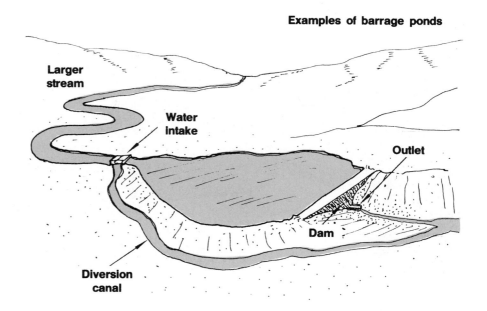

Examples of barrage ponds

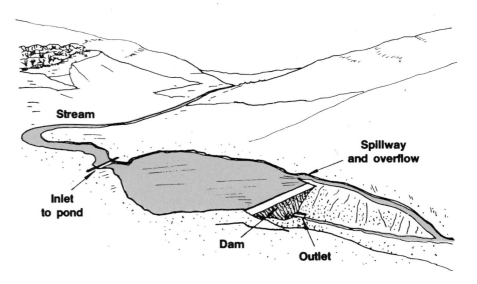

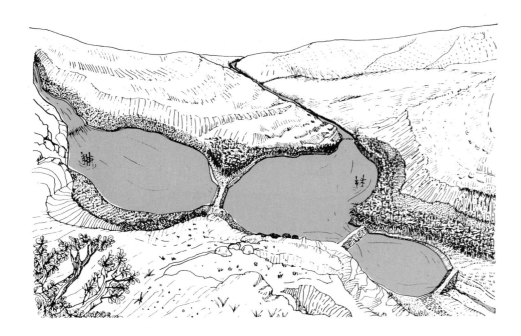

**Barrage ponds
in a V-shaped valley,
with no diversion canal**

**Barrage ponds in series
with diversion canal**

Diversion pond

10. The diversion pond is fed indirectly by gravity or by pumping through a diversion canal (which becomes the **main feeder canal**), from a spring, stream, lake or reservoir. The water flow is controlled through a water intake. There is an inlet and an outlet for each pond.

11. The diversion pond can be constructed:

- either on **sloping ground** as a cut-and-fill pond;
- or on **flat ground** as a four-dike embankment pond sometimes called a **paddy pond**.

12. It is usually drainable through a drainage canal.

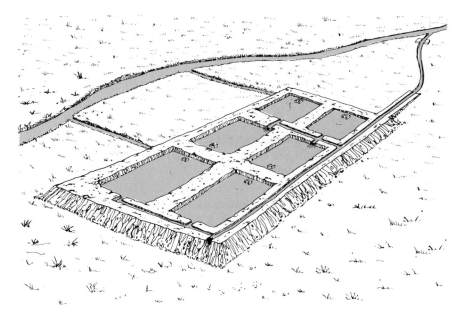

**Cut-and-fill diversion ponds
built on sloping ground**

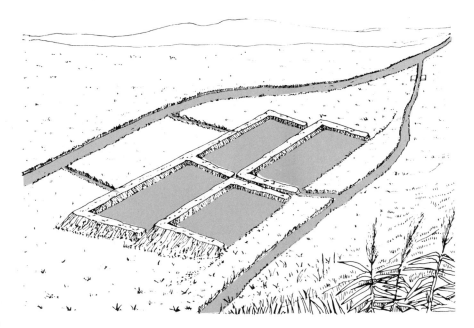

**Diversion or paddy ponds
built on flat ground
(four-dike embankment)**

15

1.   The advantages and disadvantages of the three basic types of pond that have just been defined are summarized in **Table 2**. It is important to remember the following points.

2.   Better control of the water supply means easier management of the pond, e.g. when fertilizing the water and feeding the fish.

3.   Better drainage also means easier management of the pond, e.g. when completely harvesting the farmed fish and when preparing and drying the pond bottom.

4.   A regular shape and the correct size makes a pond easier to manage and more adaptable for particular purposes.

5.   The choice of a particular type of pond will largely depend on the kind of water supply available and on the existing topography of the site selected (see Sections 16 to 18).

6.   When you have a choice of several types of pond, you should give:

- highest priority to **diversion ponds** fed by gravity;
- lowest priority to **barrage ponds** in flooding areas requiring large diversion canals.

7.   A **barrage pond** without a diversion canal should preferably be constructed only:

- to be supplied by local surface runoff and/or springs;
- across a stream with a small and regular water flow;
- below a reservoir where it will be supplied by a controlled water flow.

8.   Unless pumping is very cheap, you should not rely on it for filling or draining ponds. Do not use it where there is excessive seepage into or out of a pond.

TABLE 2

Advantages and disadvantages of the three basic types of pond

Type	Advantages	Disadvantages
Sunken pond	No need for dikes except for flood protection No water body to supply water Little skill required for construction	Water level can greatly vary seasonally Requires more work to excavate Undrainable; uncontrolled water supply, unless pumped; pumping may be expensive Low natural productivity of groundwater Pond management difficult
Barrage pond*	Simple to design for small streams Construction costs relatively low unless there are flood defence problems Natural productivity can be high, according to quality of water supply	Dike needs to be carefully anchored Need for a spillway and its drainage canal No control of incoming water supply (quantity, quality, wild fish) Cannot be completely drained except when incoming water supply dries out Pond management difficult (fertilization, feeding) as water supply is variable Irregular shape and size
Diversion pond**	Easy control of water supply Good pond management possible Construction costs higher on flat ground Can be completely drained Regular pond shape and size possible	Construction costs higher than barrage ponds Natural productivity lower, especially if built in infertile soil Construction requires good topographical surveys and detailed staking out

* If the barrage pond is built with a diversion canal, some of the disadvantages may be eliminated (controlled water supply, no spillway, complete drainage, easier pond management), but construction costs can greatly increase if the diversion of a large water flow has to be planned.
** Relative advantages will vary according to the arrangement of the ponds (see Section 16), either in series (pond management is more difficult) or in parallel (both water supply and drainage are independent, which simplifies management).

15 The physical characteristics of fish ponds

1.   Fish ponds are characterized by their **size, shape and water depth**.

Size of fish ponds

2.   The size of a fish pond is measured by its **water surface area** when the pond is full of water.

3.   The size of a barrage pond depends directly on the height of the dike built across the valley and on the topography of the valley. The length and width can be found from the longitudinal profile and from the cross-sectional profiles of the valley (see Sections 94 and 95, **Topography, 16/2**).

Area of pond = water surface at maximum water level

Length

Width

The size of a barrage pond depends on the height of the dam

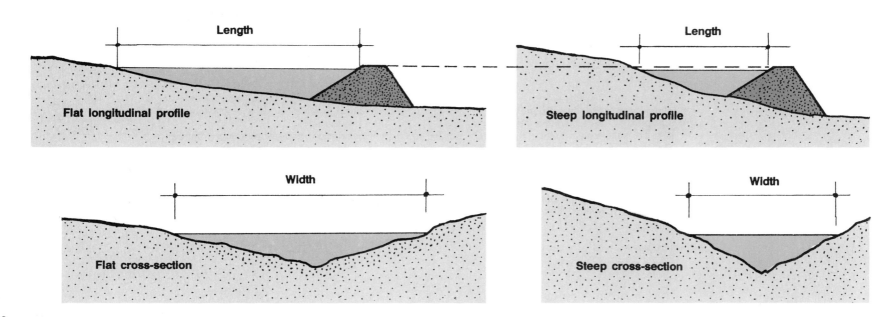

Length

Flat longitudinal profile

Length

Steep longitudinal profile

Width

Flat cross-section

Width

Steep cross-section

4. The individual size of sunken ponds and diversion ponds can be decided upon by the farmer, considering the following factors.

(a) **Use**: a spawning pond is smaller than a nursery pond, which is in turn smaller than a fattening pond.

(b) **Quantity of fish to be produced**: a subsistence pond is smaller than a small-scale commercial pond, which is in turn smaller than a large-scale commercial pond.

(c) **Level of management**: an intensive pond is smaller than a semi-intensive pond, which is in turn smaller than an extensive pond.

(d) **Availability of resources**: there is no point in making large ponds if there are not enough resources such as water, seed fish, fertilizers and/or feed to supply them.

(e) **Size of the harvests and local market demand**: large ponds, even if only partially harvested, may supply too much fish for local market demands.

**Size of fattening ponds
under semi-intensive management***
in Africa

Type of pond	Area (m^2)
Subsistence ponds	100- 400
Small-scale commercial ponds	400-1 000
Large-scale commercial ponds	1 000-5 000

* Fertilization and some feeding

**Resource availability
and pond size**

	Small pond	Large pond
Water	Small quantity rapid filling/draining	Large quantity slow filling/draining
Fish seed	Small number	Large number
Fertilizer/feed	Small amount	Large amount
Fish marketing	Small harvest Local markets	Large harvest Town markets

Note: when designing a fish farm with several fattening ponds, consider also that the **construction costs decrease as pond size increases**, and that the flexibility of **management improves as the number of ponds increases**.

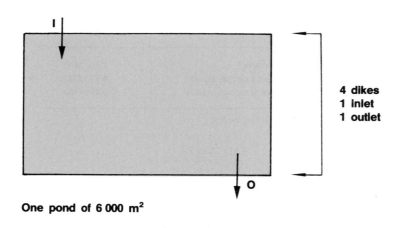

4 dikes
1 inlet
1 outlet

One pond of 6 000 m²

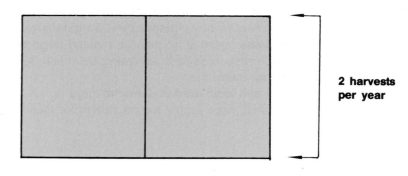

2 harvests per year

Two ponds of 3 000 m²

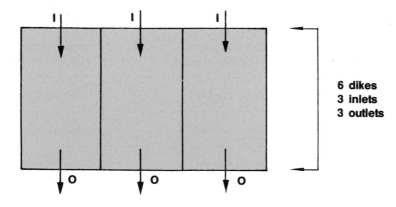

6 dikes
3 inlets
3 outlets

Three ponds of 2 000 m²

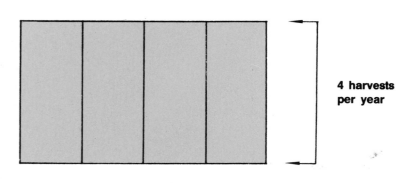

4 harvests per year

Four ponds of 1 500 m²

I = Inlet   O = Outlet

Shape of fish ponds

5. A fish pond may have any shape, as shown by barrage ponds whose shape depends exclusively on the topography of the valleys in which they are built.

6. Generally, however, sunken ponds and diversion ponds are designed with **a regular shape**, either square or rectangular. For the same pond size, the total dike length regularly increases as the pond shape gradually deviates from square and becomes more elongated. At the same time the construction costs increase.

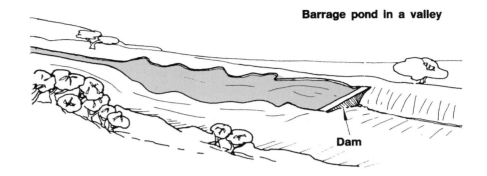

Barrage pond in a valley

Dam

Different shapes for a pond of 100 m²

Pond shape		Width (m)	Length (m)	Length of dikes (m)
Square	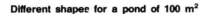	10	10	20 + 20 = 40
Rectangle		7	14.3	14 + 28.6 = 42.6
		5	20	10 + 40 = 50
		2	50	4 + 100 = 104

7. There are some cases where it may be simpler and cheaper to match the shape of the pond with the existing topography.

8. You will also find that rectangular ponds are not so much more expensive if you can build a group of them, with **shared walls** (see page 22 for diagrams comparing square and rectangular ponds).

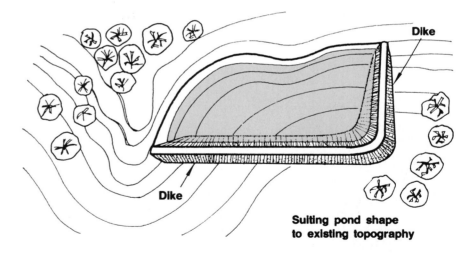

This pond makes use of existing contours and only two dikes are needed

Dike

Dike

Suiting pond shape to existing topography

Reducing the cost of pond construction by using shared dikes

SQUARE PONDS

Wall length for 1 pond = 4 × 20 m = 80 m
Wall length for 4 ponds = 12 × 20 m = 240 m

RECTANGULAR PONDS

Wall length for 1 pond = (2 × 10 m) + (2 × 40 m) = 100 m
Wall length for 4 ponds = (8 × 10 m) + (5 × 40 m) = 280 m

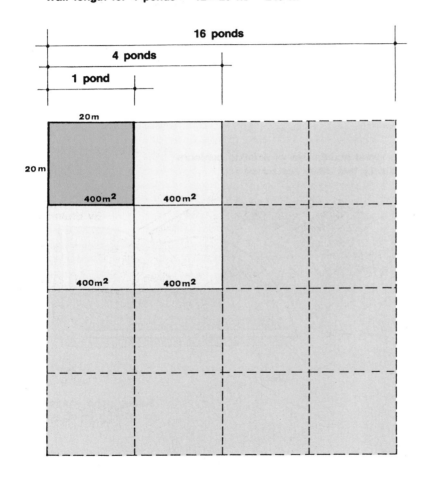

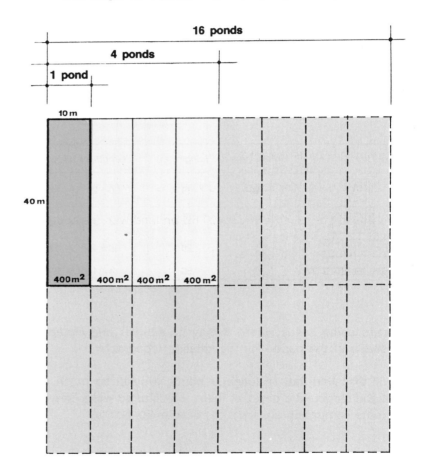

Ratio of area to wall length is a measure of relative cost
(the higher the value the better)

1 square pond	400 ÷ 80 = 5	4 square ponds 1 600 ÷ 240 = 6.7
1 rectangular pond	400 ÷ 100 = 4	4 rectangular ponds 1 600 ÷ 280 = 5.7

When square ponds are preferable

9.   Because they are cheaper to build, **square ponds** are particularly useful as smaller ponds (up to 400 m²), which you plan to harvest by draining.

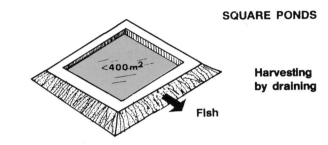

SQUARE PONDS

Harvesting by draining

<400 m²

Fish

When rectangular ponds are preferable

10.   You should prefer **rectangular ponds** whenever:

● you build ponds larger than 400 m² on land with a slope greater than 1.5 percent (see Section 17, paragraph 3);
● you build ponds larger than 100 m² and you plan to harvest your fish by seining.

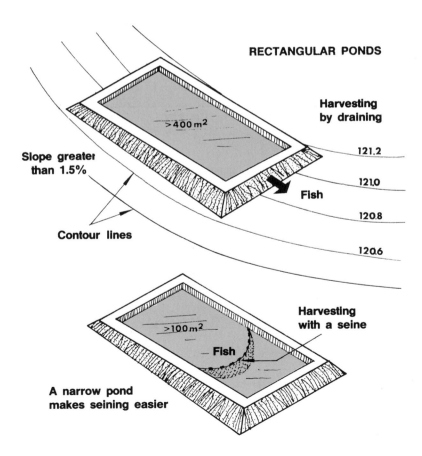

RECTANGULAR PONDS

Harvesting by draining

>400 m²

Slope greater than 1.5%

121.2

121.0

Fish

120.8

Contour lines

120.6

Harvesting with a seine

>100 m²

Fish

A narrow pond makes seining easier

Selecting a rectangular shape

11.   In general, rectangular ponds are about twice as long (**L**) as they are wide (**W**); but if you build your ponds with a bulldozer, it is cheaper to select a pond width which is a multiple of the blade width of the bulldozer.

Normal rectangular pond

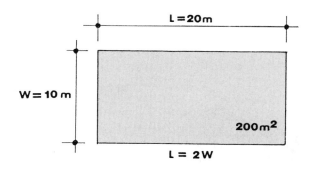

Rectangular pond excavated by bulldozer

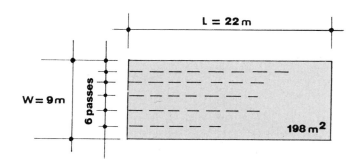

**If the bulldozer blade = 1.50 m
then the pond width should = 1.50 m × 6 (passes) = 9 m**

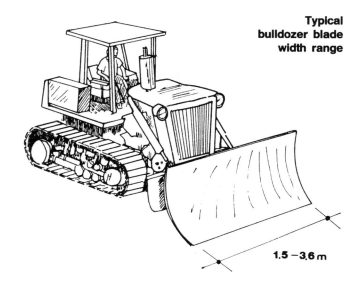

**Typical
bulldozer blade
width range**

1.5 – 3.6 m

Note: it is best to use a **standard width** for ponds that are meant for the same use. This will enable you to use standardized seine nets when harvesting them.

Type of pond	
Spawning	Width 1
Nursery 1	Width 2
Nursery 2	Width 3
Fattening	Width 4

12. Where the ground slope is greater than 1.5 percent (see Section 17), the ponds are best built with the longer sides running across the slope, with the width of the ponds limited accordingly, so that the downhill dike does not need to be too high, and so that the earth built up as walls balances the earth dug out. As the slope increases, the ponds should become narrower. You should avoid building dikes higher than three metres.

13. You should select ponds shaped to fit the local topography whenever:

- you need to use every part of the available area;
- shaping the pond this way entails good cost savings, for example by using existing earth banks or slopes;
- a regular pond shape is not too important.

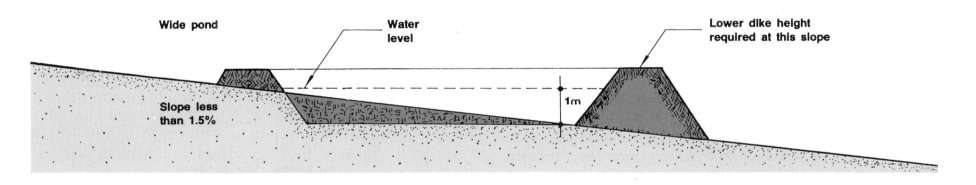

Wide pond

Water level

Lower dike height required at this slope

Slope less than 1.5%

1m

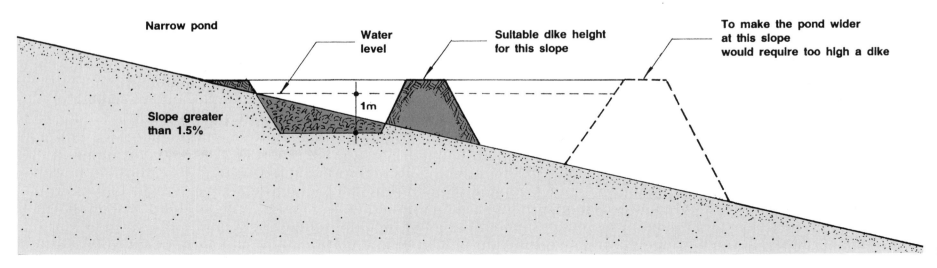

Narrow pond

Water level

Suitable dike height for this slope

To make the pond wider at this slope would require too high a dike

Slope greater than 1.5%

1m

Water depth in fish ponds

14. Except in some barrage ponds built on streams with steep longitudinal (downhill) profiles, **fish ponds are generally shallow**. Their maximum water depth does not normally exceed 1.50 m. Their shallowest area should be at least 0.50 m deep to limit the growth of aquatic plants. The water depth in small rural ponds normally varies from 0.50 m (shallow area) to 1 m at the most (deep area).

15. Deeper ponds are much more expensive to build, because the volume of the dikes increases rapidly as you make ponds deeper.

16. Sometimes it is necessary, however, to use deeper ponds:

*   **in dry regions** where you need to store water through the dry season to make sure there is enough for the fish;
*   **in cold regions** where it may be necessary to provide the fish with a refuge in deeper, warmer waters during cold weather.

Note: during the cold season, it is sometimes better to keep the main ponds dry and to hold the fish in smaller, deeper wintering ponds. In such cases, the main ponds can be designed more cheaply. They will also warm up more quickly than deep ponds in spring.

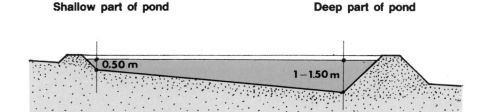

Shallow part of pond   **Deep part of pond**

0.50 m   1 – 1.50 m

**Dike volume per metre
of dike length
(see Section 64)**

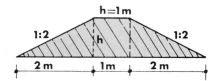

h=1 m

1:2   1:2

2 m   1m   2 m

Dike volume = 3 m³ per metre

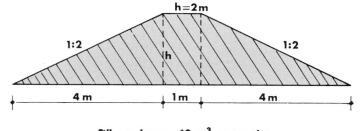

h=2 m

1:2   1:2

4 m   1m   4 m

Dike volume = 10 m³ per metre

Characteristics of shallow and deep ponds

Shallow ponds	Deep ponds
Water warms up rapidly	Deep water warmer in cold season
Great fluctuations of temperature	Water temperature more stable
Greater danger from predatory birds	Less natural food available
Greater growth of water plants	Difficult to seine in deep water
Smaller dikes needed	Strong, high dikes needed

16 How to select the pond to suit local topography

1. In the previous manual of this series (see Section 82, **Topography**, **16/2**), you learned how to make a cross-section profile of a valley. From the general shape of this profile, you can already decide upon the type of pond which could be built:

- if the valley is deep, steep and narrow, do not build ponds;
- if the valley bottom is 50 to 100 m wide, barrage ponds might be appropriate;
- if the valley bottom is more than 100 m wide, diversion ponds could be built.

2. A more detailed study should confirm your choice, based on the longitudinal profile and on the cross-section profile of the valley. Select the type of ponds to build:

- either according to **the shape** of the valley and its profiles (see **Table 3**);
- or according to **the slope** of the longitudinal profile (downhill) and the cross-section profile of the valley (see **Table 4**).

Valley unsuitable for pond building

Valley suitable for pond building

TABLE 3

Selection of pond type according to shape of valley

Type of pond	Shape of valley cross-section profile			
	V	Rounded V	Centrally truncated V	Laterally truncated V
Sunken pond	—	Whenever groundwater (spring or seepage) or runoff is available		
Barrage pond	If longitudinal profile of valley has slope less than 5%	—	—	If longitudinal profile has slope less than 5% and cross-section profile has slope 5-10%
Diversion pond: cut-and-fill type	—	Where cross-section profile has slope less than 5%	Where cross-section profile has slope 0.5-5%	
Diversion pond: paddy type	—	—	Where cross-section profile has slope less than 0.5%	

TABLE 4

Selection of pond type according to valley slope

Valley longitudinal profile (downhill)	Valley cross-sectional profile	Possible type of pond
Slope greater than 5%	Slope greater than 5%	None
	Slope less than 5%	Diversion pond Sunken pond
Slope less than 5%	Slope 5-10%	Barrage pond Sunken pond
	Slope less than 5%	Diversion pond Paddy pond Sunken pond

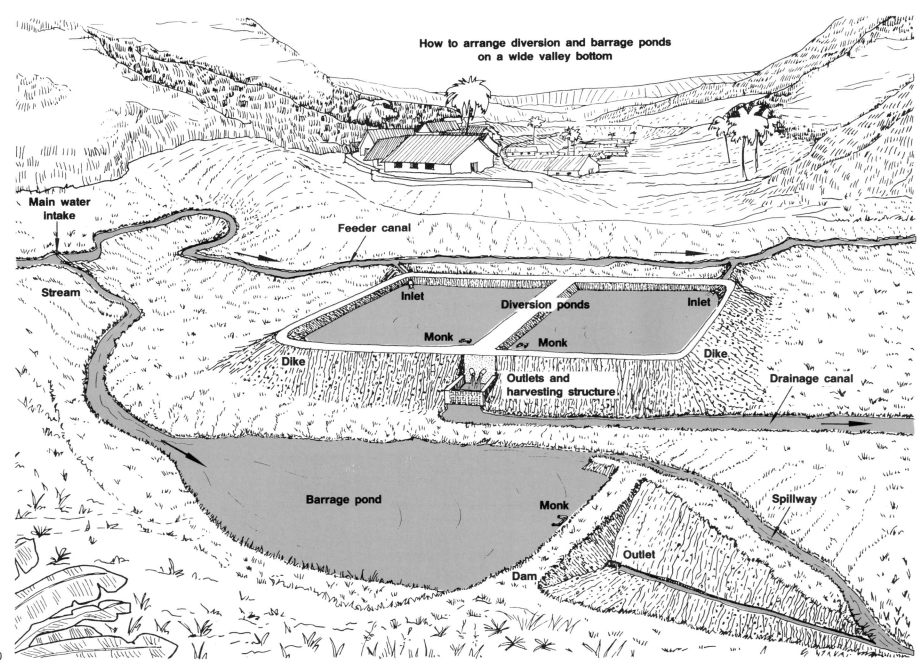

How to arrange diversion and barrage ponds on a wide valley bottom

Main water intake

Feeder canal

Stream

Inlet

Diversion ponds

Inlet

Monk

Monk

Dike

Dike

Outlets and harvesting structure

Drainage canal

Barrage pond

Monk

Spillway

Outlet

Dam

17 Laying out fish ponds

1. You have already learned (see Section 12) that several structures may be required for the good functioning of your fish ponds, particularly if you plan to have several of them in production. In the next book of this manual, **Pond construction**, **20/2**, you will learn how to build various structures, but right now it is important to understand the different possibilities which exist for the layout of your ponds and their structures.

2. It will always be easier to lay out your ponds if the land you select slopes slightly and if you can supply water along its highest contour line, i.e. at the top end of the site.

Pond and pond structures

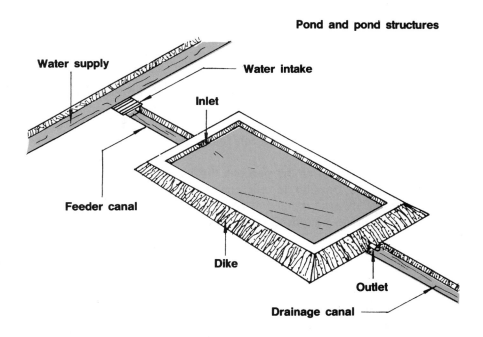

Pond on sloping ground

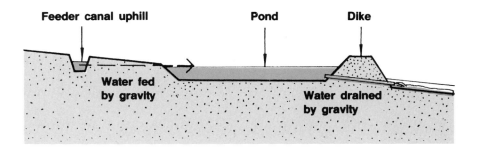

Ponds on a slope

3. If diversion ponds are built on a slope, their orientation should vary according to the angle of slope so that there is a minimum amount of earthwork:

- **slope 0.5-1.5 percent:** the length of rectangular ponds should be perpendicular to the **contour lines***, i.e. the ponds should run downhill so that the floor of the pond will follow the natural slope, and no excavation will be needed to make the deeper part of the ponds;
- **slope greater than 1.5 percent:** the length of rectangular ponds should be parallel to the **contour lines***, i.e. the ponds run perpendicular to the slope. You would make the ponds narrower as the slope increases (see Section 15, paragraph 10).

Wider ponds

Slope 3 percent = 3 cm/m (maximum 30 m width, parallel to contour)

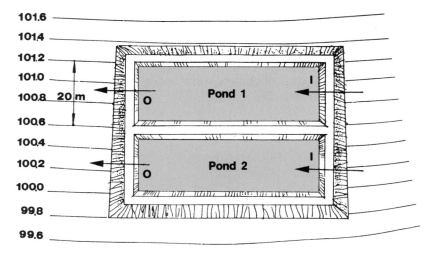

Slope 1 percent = 1 cm/m (perpendicular to contour)

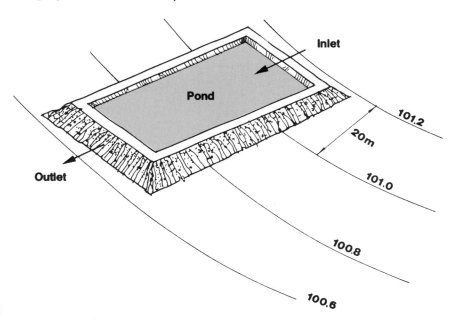

Narrower ponds

Slope 5 percent = 5 cm/m (maximum 20 m width, parallel to contour)

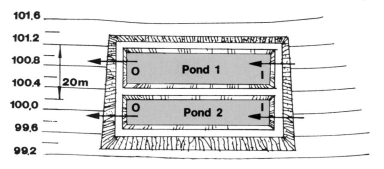

I = Inlet   O = Outlet

32

4. If there is more than one pond, you should try to share structures such as dikes, feeder canals or drains. To reduce costs, keep the length of the canals and drains as short as possible.

Ponds with shared structures

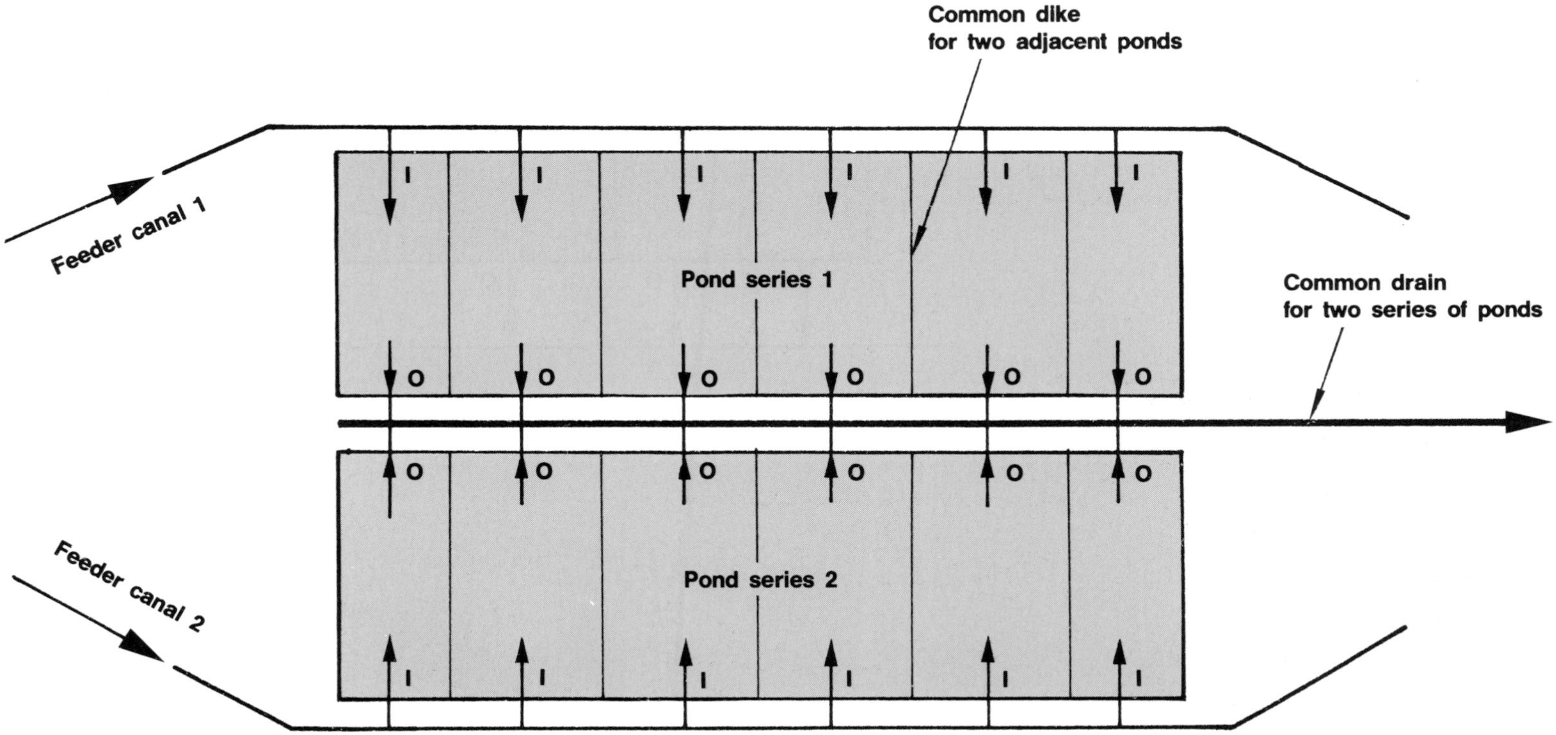
Common dike
for two adjacent ponds
Feeder canal 1
I
Pond series 1
O
Common drain
for two series of ponds
Pond series 2
Feeder canal 2

I = Inlet O = Outlet

5. Lay out your fish ponds in one of the following ways.

(a) **In series:** ponds depend on each other for their water supply, the water running from the upper ponds to the lower ponds.

(b) **In parallel:** ponds are independent from each other, each pond being supplied directly from the feeder canal. Water has not been used after passing through another pond. This layout is to be preferred.

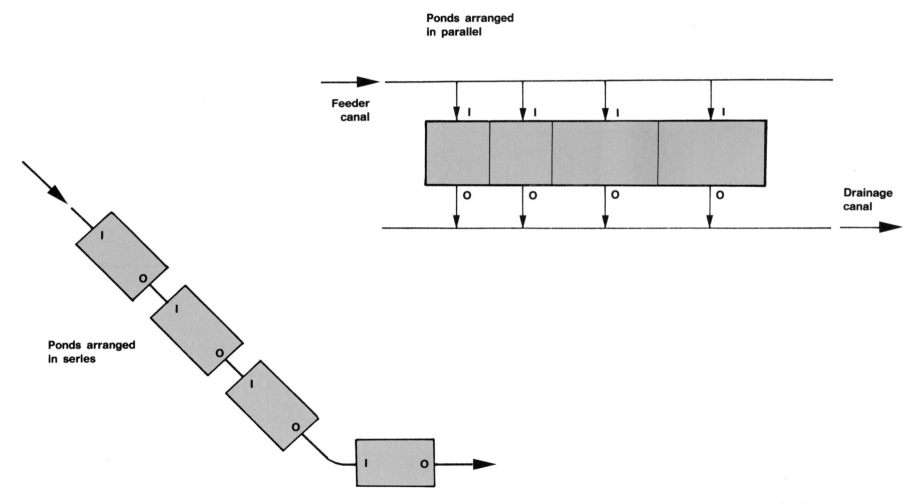

Ponds arranged in parallel

Feeder canal

Drainage canal

Ponds arranged in series

I = Inlet   O = Outlet

6. It is always best to provide **a means of diverting excess water.** In the case of barrage ponds, a diversion canal can carry the water around the pond to a point downstream from the barrage. For diversion ponds, the excess water is simply allowed to flow in the natural stream bed instead of in the feeder canal.

7. All feeder canals should end in a drain, so that any excess water in the canal can be discharged away from the ponds.

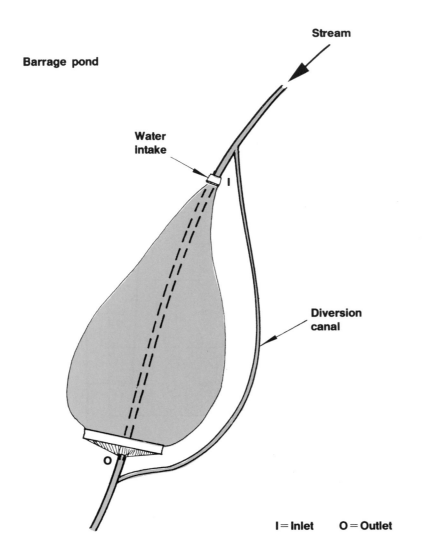

Barrage pond

Stream

Water Intake

Diversion canal

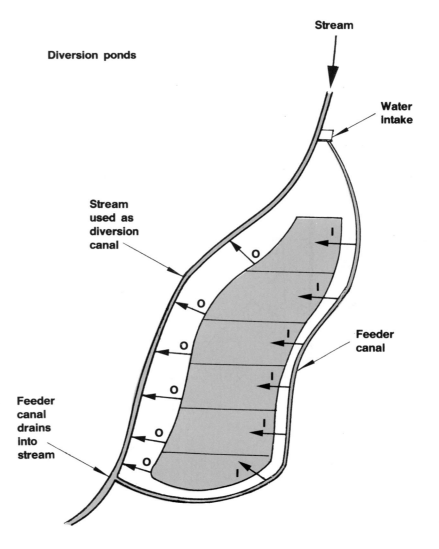

Diversion ponds

Stream

Water Intake

Stream used as diversion canal

Feeder canal

Feeder canal drains into stream

I = Inlet   O = Outlet

8. With **several rows of ponds,** it is always best to arrange feeder and drainage canals to serve a row of ponds on both sides of the canals.

Note: on the following pages are a number of examples showing the layout of various kinds of fish ponds.

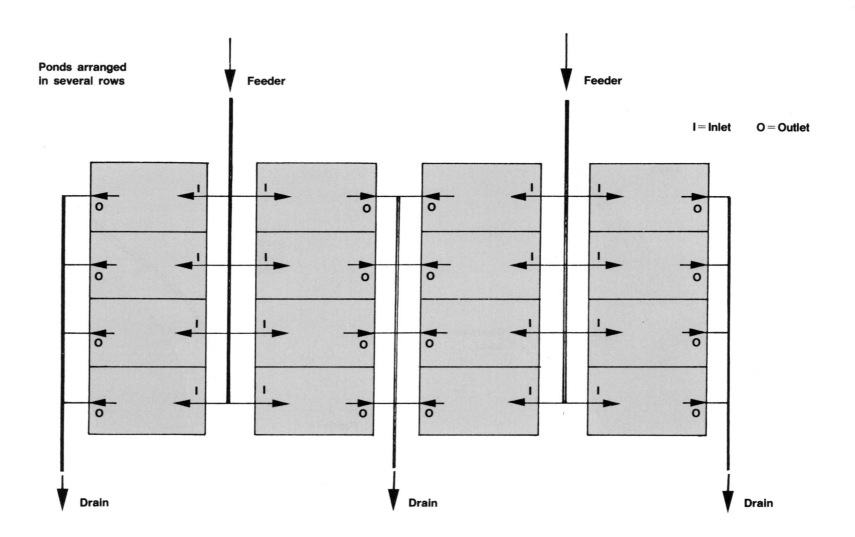

Ponds arranged in several rows

Feeder

Feeder

I = Inlet   O = Outlet

Drain

Drain

Drain

Sunken ponds
- Fed by seepage
- Built next to one another at bottom of a valley

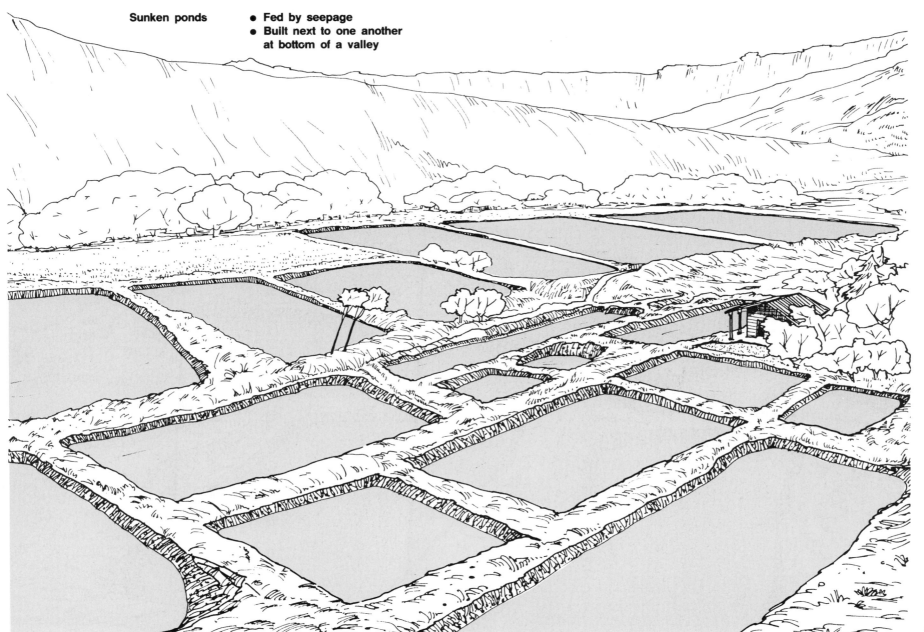

Sunken ponds
- Fed by seepage
- Built next to one another at bottom of a valley

Barrage ponds
- Stream fed
- Arranged in series with diversion canal
- Impractical and expensive if there are large floods

Stream

Diversion canal

Barrage ponds

- Spring fed
- Arranged in series without diversion canal
- Management rather difficult

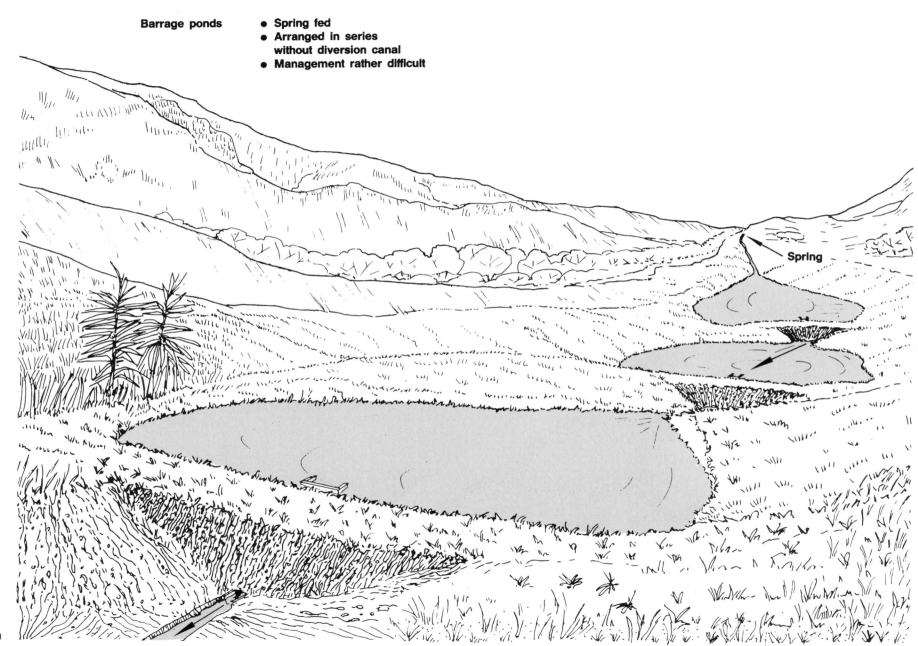

Spring

Diversion ponds
- Stream fed
- One line of ponds
- Arranged in parallel with natural diversion canal
- Best water control
- Higher investment than ponds arranged in series

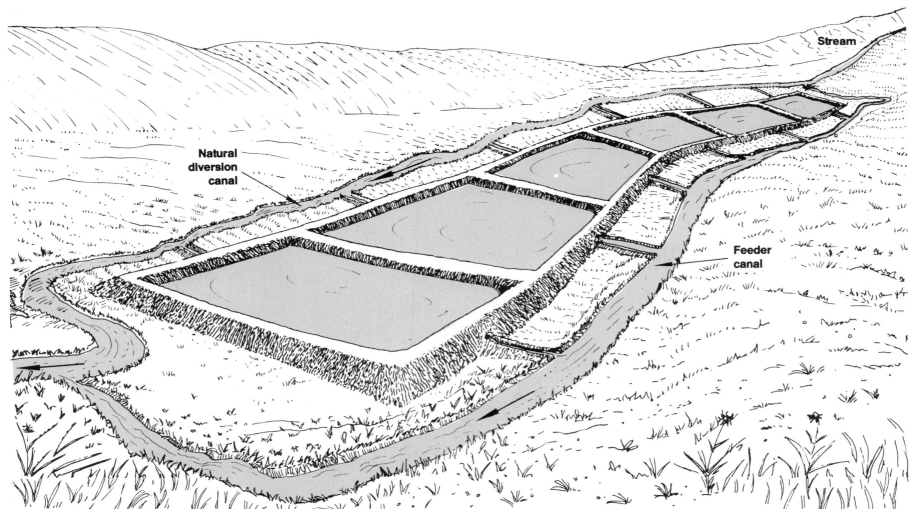

Stream

Natural diversion canal

Feeder canal

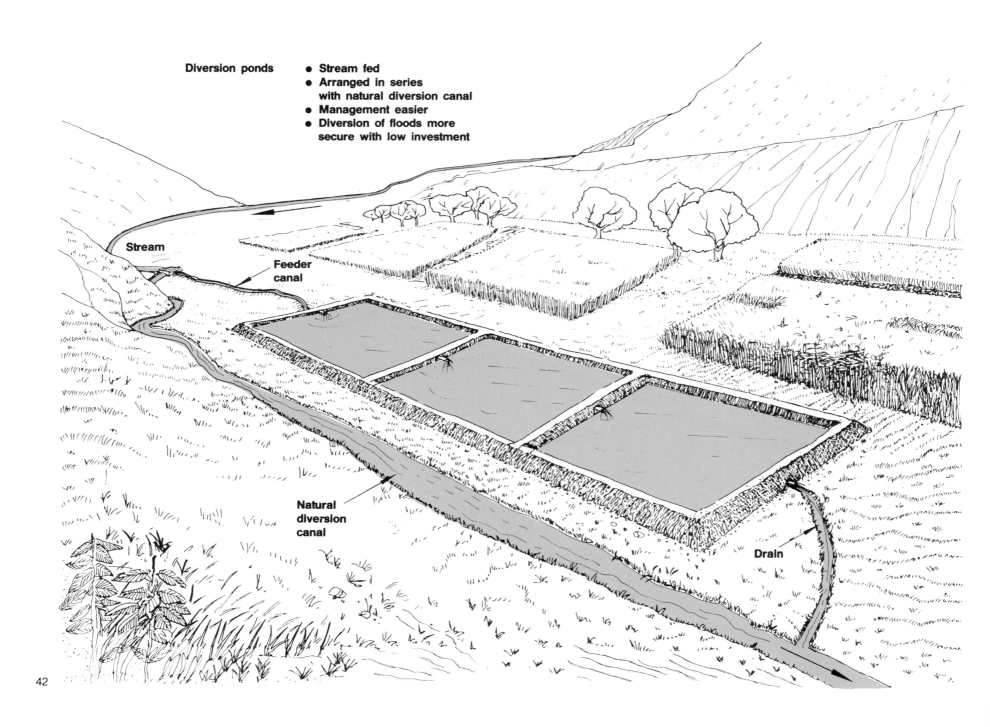

Diversion ponds
- Stream fed
- Arranged in series with natural diversion canal
- Management easier
- Diversion of floods more secure with low investment

Stream

Feeder canal

Natural diversion canal

Drain

42

Diversion ponds
- Fed from reservoir
- Two lines of ponds
- Arranged in parallel and rows with central natural diversion canal
- Commercial size farm

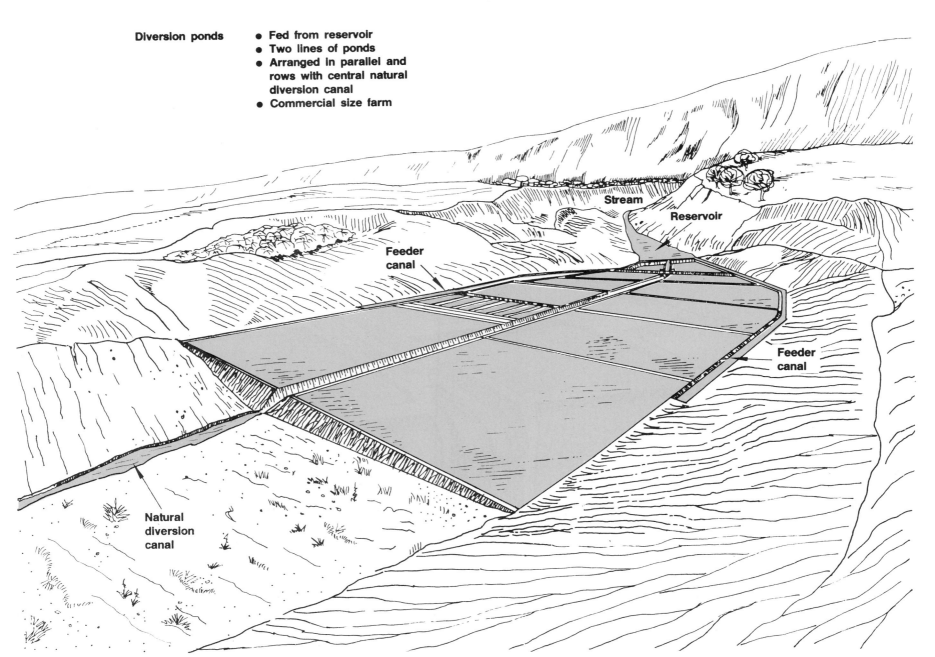

Stream

Reservoir

Feeder canal

Feeder canal

Natural diversion canal

43

Diversion ponds
- Stream fed
- Two lines of ponds
- Arranged in parallel with natural diversion canal
- Best water control
- Higher investment than ponds arranged in series

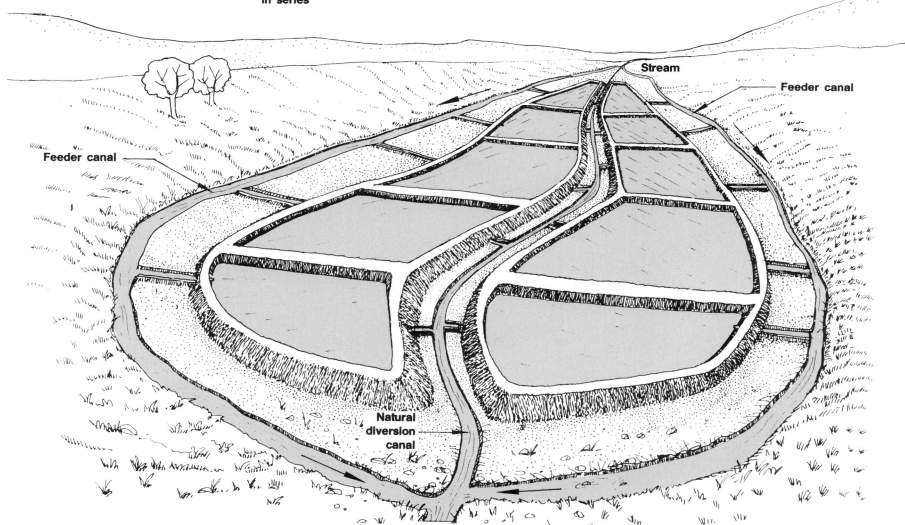

Stream

Feeder canal

Feeder canal

Natural diversion canal

Barrage ponds below a reservoir

- **Fed by reservoir seepage or overflowing water**
- **Arranged in series**

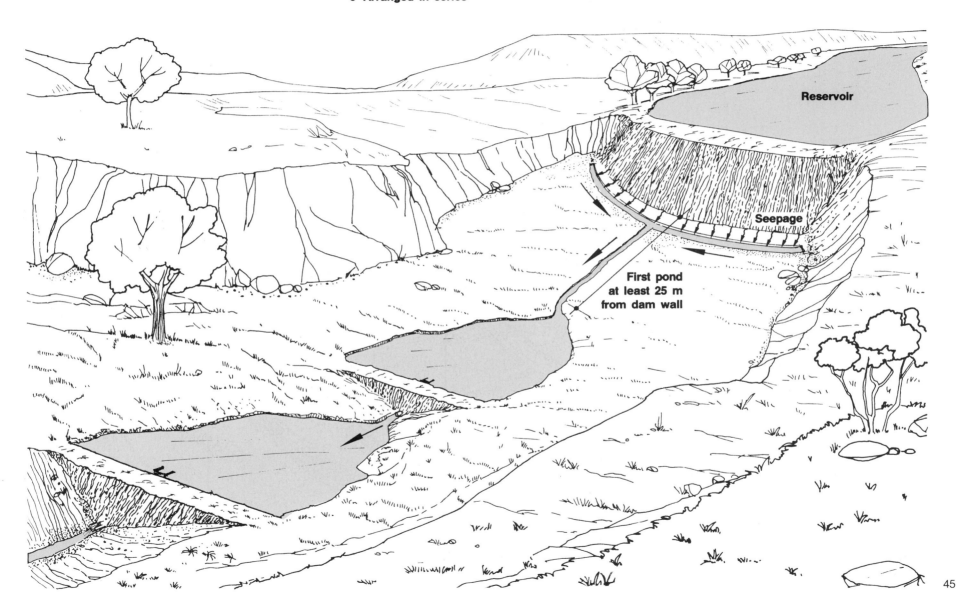

Reservoir

Seepage

First pond at least 25 m from dam wall

45

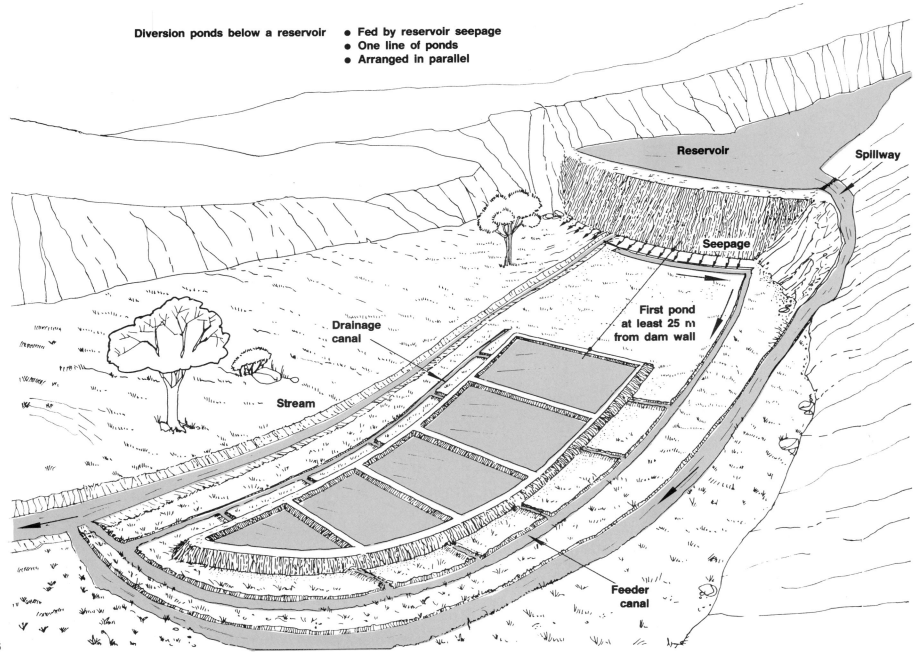

Diversion ponds below a reservoir
- Fed by reservoir seepage
- One line of ponds
- Arranged in parallel

Reservoir

Spillway

Seepage

First pond at least 25 m from dam wall

Drainage canal

Stream

Feeder canal

46

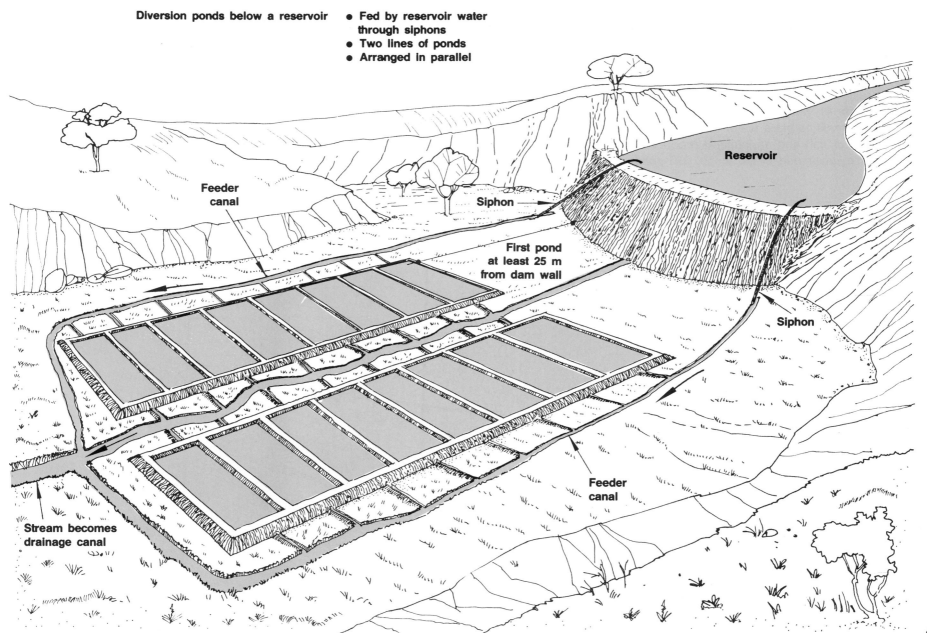

Diversion ponds below a reservoir

- Fed by reservoir water through siphons
- Two lines of ponds
- Arranged in parallel

Feeder canal

Siphon

Reservoir

First pond at least 25 m from dam wall

Siphon

Feeder canal

Stream becomes drainage canal

Integrated farming

9. The production of fish in ponds can easily be integrated with agricultural production, particularly on sloping ground.

(a) On the slope itself, trees may produce wood, fuel and food. The forest cover will protect the soil well and control erosion (see Section 41, **Management**, **21/1**).

(b) A reservoir to store water during the dry season may be built and used for fish and agricultural production.

(c) At the lower end of the slope, fish ponds can be built.

(d) Various kinds of animals can be raised next to these ponds and can provide fertilizer for them (see Chapter 7, **Management**, **21/1**).

(e) Water from the ponds may be used for watering adjacent gardens and crops.

(f) Mud that accumulates on the bottom of the ponds can periodically be removed to fertilize surrounding crops.

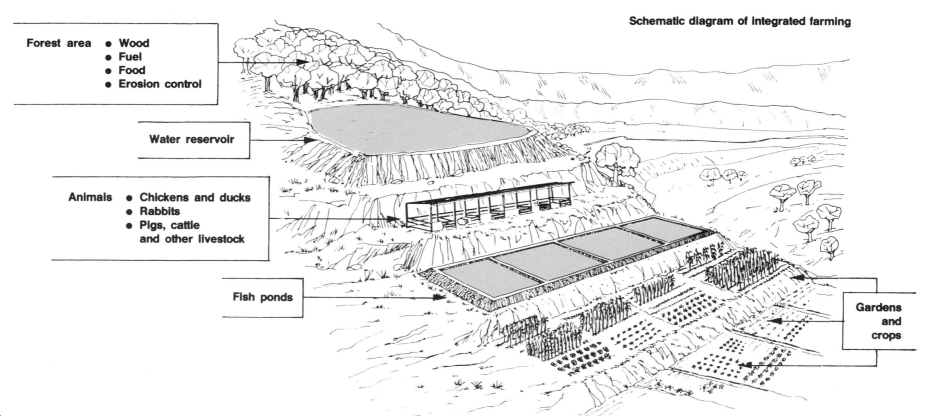

Schematic diagram of integrated farming

Forest area
- Wood
- Fuel
- Food
- Erosion control

Water reservoir

Animals
- Chickens and ducks
- Rabbits
- Pigs, cattle and other livestock

Fish ponds

Gardens and crops

Integrated fish farming

10.   On your fish farm you may combine two production systems in two separate groups of ponds:

- **an intensive system** where fish are densely stocked and where their growth is sustained by adequate feeding, using fertilizers and feeds (see **Management, 21/2**);

- **an extensive system** where fish are stocked at a lower density and where their growth relies only on the presence of natural food; the production of this natural food is enhanced by the rich water draining out of the intensive system into the extensive one.

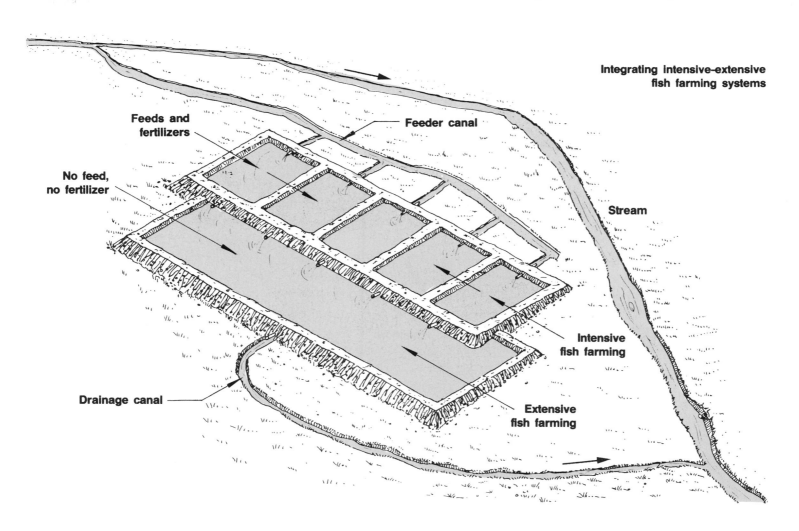

Integrating intensive-extensive fish farming systems

Feeds and fertilizers

Feeder canal

No feed, no fertilizer

Stream

Intensive fish farming

Drainage canal

Extensive fish farming

**Integrated fish farming
and crop production
in a swampy area**

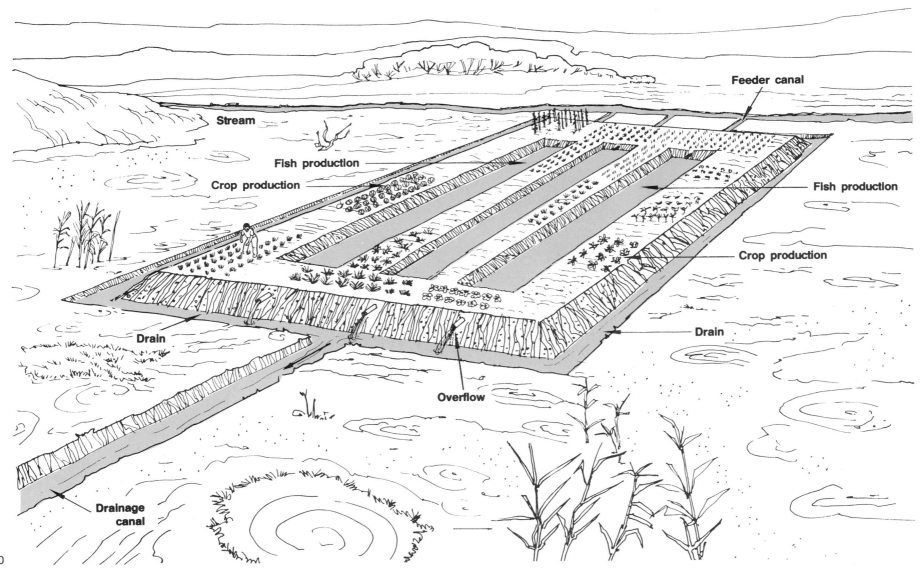

Stream

Feeder canal

Fish production

Crop production

Fish production

Crop production

Drain

Drain

Overflow

Drainage
canal

50

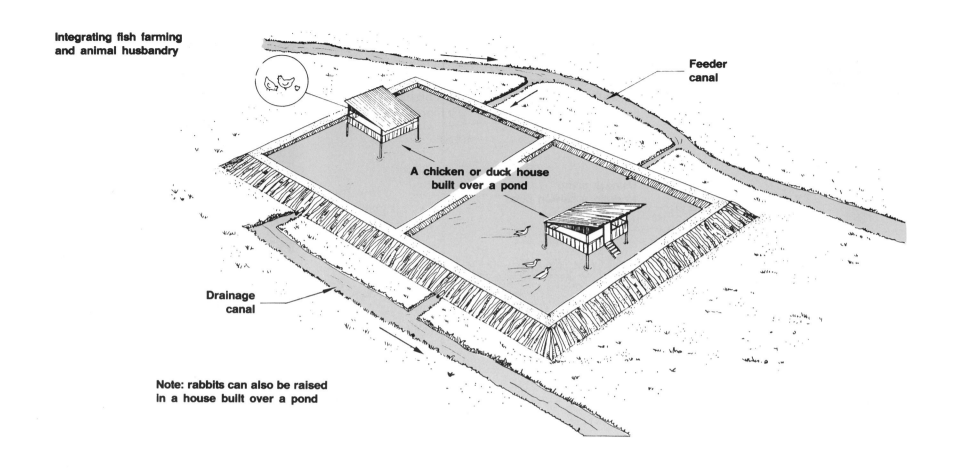

Integrating fish farming and animal husbandry

Feeder canal

A chicken or duck house built over a pond

Drainage canal

Note: rabbits can also be raised in a house built over a pond

Chicken or duck house

Chicken or duck house

51

11.   Pumping is not normally used in those layouts fed from streams or reservoirs but can be used for sunken ponds and sometimes to supply diversion ponds from a lake or reservoir. In times of severe water shortage, pumps can be used to recycle the waste water, drawn from the drainage canals and fed back to the feeder canals. By using pumps where manual methods would be limited, you can sometimes take advantage of sites or plan your ponds more flexibly. However, using a pump involves additional costs, and re-using the waste water may cause problems to the fish. Recycling should only be considered in an emergency.

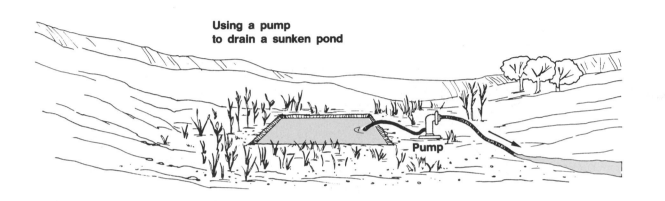

**Using a pump
to drain a sunken pond**

**Note:
in some cases
drainage water may be
returned to the supply**

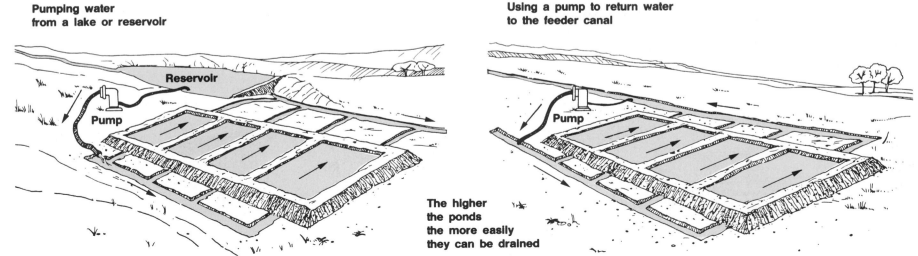

**Pumping water
from a lake or reservoir**

Reservoir

Pump

**Using a pump to return water
to the feeder canal**

Pump

**The higher
the ponds
the more easily
they can be drained**

18 How to plan your fish farm

Considering its size and complexity

1. The size of a fish farm will vary according to the level of production you wish to reach. The greater the potential fish production, the greater the investment, and the greater the farm size. The number and size of ponds increase as the fish farm increases in scale. The culture system also becomes more complete, with special ponds for broodstock, fry and fingerlings, and storage, as well as the main ponds for producing food fish.

2. **Subsistence fish farmers** do not need more than one or two small ponds, which are used as fattening ponds and sometimes also as breeding/nursery ponds (**culture system A**). This system can be improved by adding one or more small storage ponds to keep the harvested juveniles alive while the fattening pond is harvested, repaired and refilled with water (**culture system B**).

3. **Small-scale commercial fish farms** usually add one or more spawning ponds and nursery ponds, making the farm independent as far as seeds are concerned (**culture systems C and D**). Pond numbers and sizes slightly increase. One or more storage ponds can also be used for marketing.

4. **Large-scale commercial fish farms** may have the most complete range of fish-rearing facilities, including brood ponds and nursery ponds (**culture systems D and E**). Pond numbers and sizes greatly increase.

5. As the fish-rearing facilities increase in size and become more complex, other facilities (the support infrastructure) also become important. These may include roads, power production and distribution, feeds production and storage, workshops, office/laboratory, hatchery, housing, etc.

**Change of the cultural system
from subsistence to commercial scale**

Pond type	Cultural system				
	A	B	C	D	E
Brood					●
Spawning			●	●	●
Nursery 1 (fry)				●	●
Nursery 2 (fingerlings)				●	●
Fattening	●	●	●	●	●
Storage		Juvenile stocking	Juvenile stocking		
Marketing				●	●
Production level	Subsistence		Small-scale commercial		Large-scale commercial

6. The layout becomes more difficult to design as the fish farm grows in size and complexity. Remember, the design of large farms is best done by a specialized engineer. However, to lay out a smaller-scale farm, the following guidelines will be useful.

Laying out ponds according to their use

7. Ideally, the entire pond area should be visible from the main office/service building area at the centre of the farm. For very big farms, it may be necessary to group the ponds, each with its own small working centre.

8. Lay out the brood ponds, spawning ponds and storage ponds so that they are well protected against poaching, easily accessible by vehicle, easily drainable and well supplied with good quality water.

9. Lay out the nursery ponds between the spawning ponds and the fattening ponds. Provide easy access for at least a mini-tractor and its trailer.

10. Lay out the fattening ponds to allow good access for feeds, fertilizers and equipment as well as easy transfer of harvested fish to storage ponds or the outside market.

Laying out the access roads on your farm

11. To have better control over incoming and outgoing traffic, restrict access to the farm to one point only. It is sometimes preferable to group most of the service buildings near this access point.

12. Limit the canal crossings to the minimum.

13. Build road crossings on the feeder canals rather than on drainage canals, as these are usually narrower and easier to cross. This might require keeping the main access road along the higher side of the farm.

14. Provide access as near as possible to the harvesting area of the fish ponds. By grouping harvesting areas together, a single access point can serve several ponds.

15. Have good access on the farm itself to the main water control structures. Try to make sure they are all within the farm boundary.

16. Design access roads and their turning points according to the particular type of vehicle you plan to use on them: the narrower the road, the cheaper it is to build and maintain (see also Section 61).

Laying out the canals on your farm

17. Try to make each canal serve ponds on both sides.

18. Try to minimize the total length of canals, unless it makes laying out the ponds too difficult or their construction too expensive.

19. Try to make canal networks reasonably straight and simple. Minimize the number of junctions.

20. Try to avoid drainage and feeder canals that have to cross each other.

21. Try to avoid canals that have to run down a slope steeper than 5 percent.

LAYOUT FOR A COMMERCIAL FISH FARM

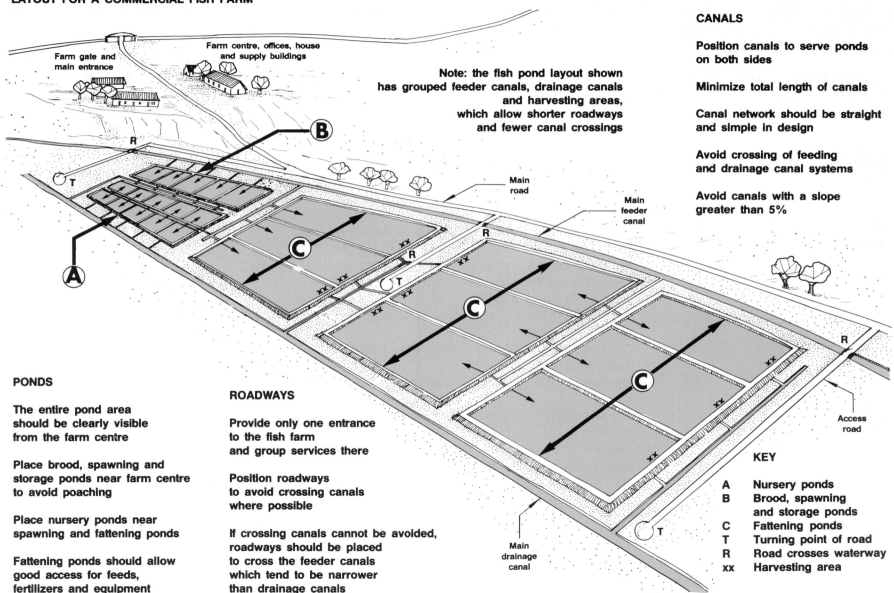

Farm gate and main entrance

Farm centre, offices, house and supply buildings

Note: the fish pond layout shown has grouped feeder canals, drainage canals and harvesting areas, which allow shorter roadways and fewer canal crossings

Main road

Main feeder canal

Access road

Main drainage canal

CANALS

Position canals to serve ponds on both sides

Minimize total length of canals

Canal network should be straight and simple in design

Avoid crossing of feeding and drainage canal systems

Avoid canals with a slope greater than 5%

PONDS

The entire pond area should be clearly visible from the farm centre

Place brood, spawning and storage ponds near farm centre to avoid poaching

Place nursery ponds near spawning and fattening ponds

Fattening ponds should allow good access for feeds, fertilizers and equipment

ROADWAYS

Provide only one entrance to the fish farm and group services there

Position roadways to avoid crossing canals where possible

If crossing canals cannot be avoided, roadways should be placed to cross the feeder canals which tend to be narrower than drainage canals

KEY

A	Nursery ponds
B	Brood, spawning and storage ponds
C	Fattening ponds
T	Turning point of road
R	Road crosses waterway
xx	Harvesting area

22. When laying out your fish farm and, later, when designing your fish ponds, it is important that you clearly understand how the elevation of the various structures has to change progressively to ensure a gravity* water flow.

23. If you plan to have either **barrage ponds** or **diversion ponds** fed by gravity, remember:

(a) Water flows down from the highest to the lowest point.
(b) The water surface in a pond is always horizontal.
(c) The pond bottom should be above the water table at harvest.
(d) The bottom of the main water intake should be below the minimum level of the water source.
(e) The bottom of the feeder canal should be at or above the maximum pond water level.
(f) The pond inlet should be located at or above the maximum pond water level.
(g) The start of the pond outlet should be at the lowest point of the pond.
(h) The end of the pond outlet should be at or above the water level in the drain.
(i) The end of the drain should be at or above the maximum water level in the natural channel.

a Water flows down by gravity

b Water surface horizontal

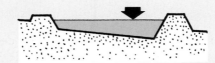

c Pond bottom above watertable at harvest

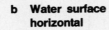

d Bottom of main water intake below water source

Source

Intake

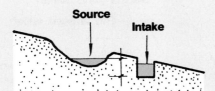

e Bottom of feeder canal at or above maximum pond water level

Feeder canal

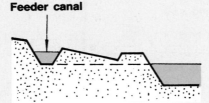

f Pond inlet at or above maximum pond water level

Inlet

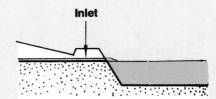

g Start pond outlet at lowest point of pond

Outlet

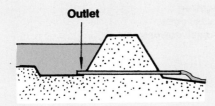

h End of pond outlet at or above water level in drain

Drain

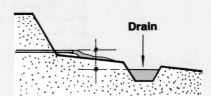

i End of drain at or above maximum channel water level

Drain

Channel

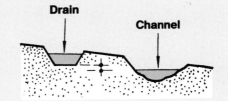

If you are building a barrage pond

24. In the case of a barrage pond fed directly by a small stream, it is easy to determine the difference in level (**X**) required between the **maximum water level upstream** and the **maximum water level downstream** from the pond that will provide enough depth of water in the barrage pond: **X** should be at least 0.80 m.

Barrage pond level differences

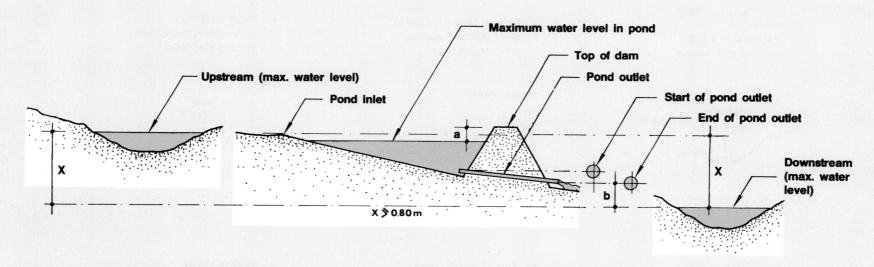

X = **The difference in level required between the maximum water level upstream and the maximum water level downstream**

a = **The difference in level required between the top of the dam and the maximum water level in the pond (freeboard)**

b = **The difference in level required between the end of the pond outlet and the maximum water level downstream**

If you are building a diversion pond

25. In the case of a diversion pond fed from a stream through a main water intake and a feeder canal, it is easy to determine the difference in level (**X**) required between **minimum water level at the main intake** and **maximum water level at the end of the drain**: **X** should be at least 1.20 m.

Diversion pond level differences

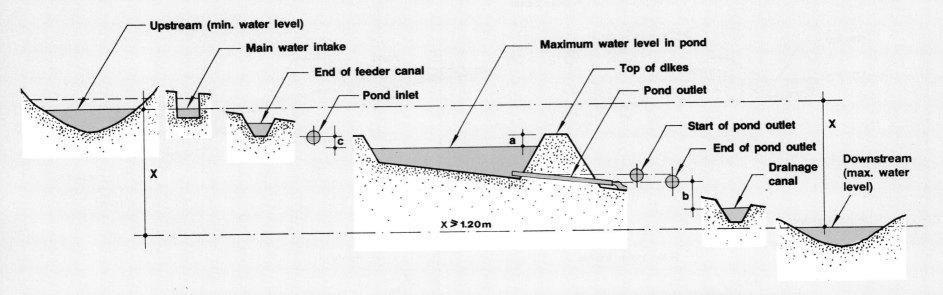

Upstream (min. water level)

Main water intake

End of feeder canal

Pond inlet

Maximum water level in pond

Top of dikes

Pond outlet

Start of pond outlet

End of pond outlet

Drainage canal

Downstream (max. water level)

X

X ⩾ 1.20m

X = The difference in level required
between the minimum water level at the main intake
and the maximum water level at the end of the drainage canal

a = The difference in level required
between the top of the dikes
and the maximum water level in the pond

b = The difference in level required
between the end of the pond outlet
and the maximum water level in the drainage canal

c = The difference in level required
between the pond inlet
and the maximum water level in the pond

59

A pump might be necessary

26.   If the topography of the site does not allow you to create these differences in level, and you can afford a pump, it is sometimes possible to pump water up from a stream into a feeder canal, or more often, to pump the last 10-20 cm of water out from a draining pond. In these cases, you can reduce the values of **X**, but you must be sure that the cost of pumping is acceptable.

Pond filling

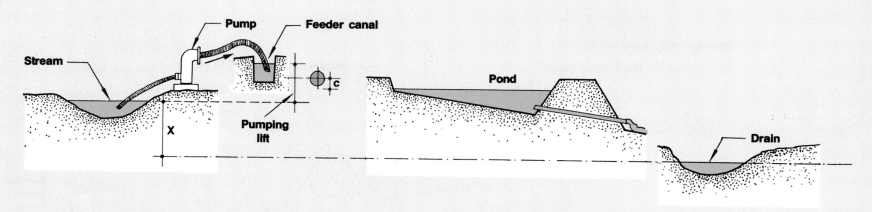

Pond draining

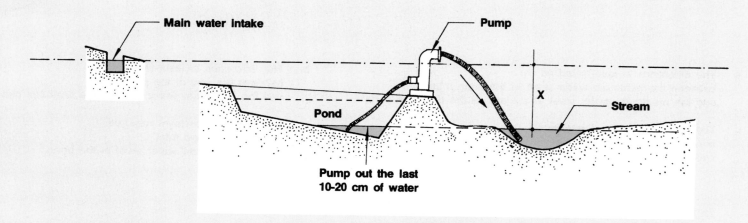

Living on your fish farm

27.   It is always desirable that somebody lives on your fish farm next to your ponds, not only for security reasons but also to be able to manage the farm properly.

28.   If the fish farm is built on sloping land, it is best to site the housing at a higher level, so that ponds can be observed from there.

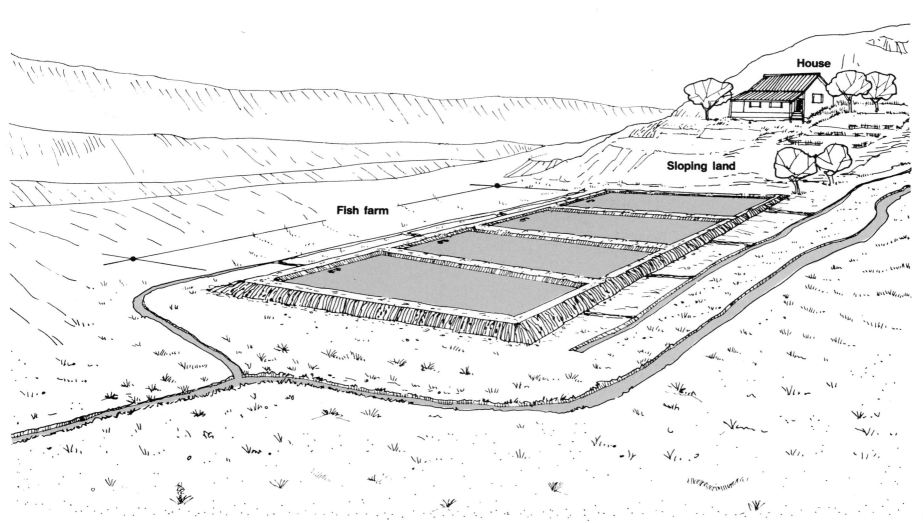

29. If the fish farm is built on flat land, you may need to site the housing on a raised platform served by a road; such a platform could also be used for storage of equipment or feeds or fertilizers, and for small animal husbandry.

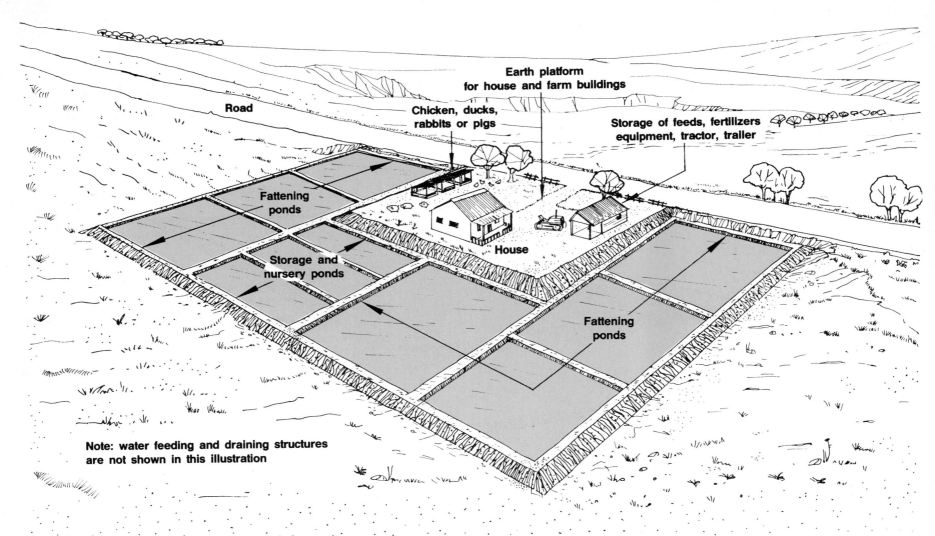

Road

Earth platform
for house and farm buildings

Chicken, ducks,
rabbits or pigs

Storage of feeds, fertilizers
equipment, tractor, trailer

Fattening
ponds

Storage and
nursery ponds

House

Fattening
ponds

**Note: water feeding and draining structures
are not shown in this illustration**

2   SITE SELECTION AND GENERAL PLANNING

20 The importance of a good site

1. The whole future success of your enterprise depends on the selection of a good site for your fish farm. The layout and the management of your farm will largely be influenced by the kind of site you select. The site therefore will strongly affect the cost of construction, the ease with which the ponds can be managed, the amount of fish produced and, in general, the economics of your enterprise.

The selected site

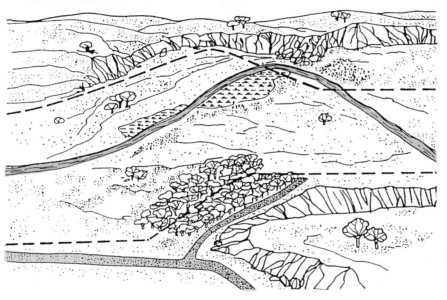

21 Preliminary decisions

1. Before starting to look for a site, you should have a clear idea of the type of fish farm you wish to build. Some of the questions you should ask yourself are the following.

(a) Which level of production do I plan to reach, subsistence or commercial? Which scale, if commercial?
(b) Which culture system shall I adopt?

● extensive or intensive;
● one or several species;
● seasonal or year-round.

(c) Shall I use fertilizers or fish feeds or both?
(d) Which species of fish shall I produce and at which size shall I sell them?
(e) Shall I have to buy juveniles or produce them all myself?
(f) Shall I try to integrate my fish farming activities with other agricultural productions I already have? Shall I also start raising animals on my farm?
(g) Which area of the farm do I wish to develop immediately? Will I develop other areas later, as a second phase?

2. If you cannot answer these questions by yourself, you should look for assistance, for example from the local extension agent specialized in fish farming. You can also check with other farmers to find out what choices they made and why.

HOW TO SELECT A SITE FOR FRESHWATER FISH CULTURE

The decision-making cycle

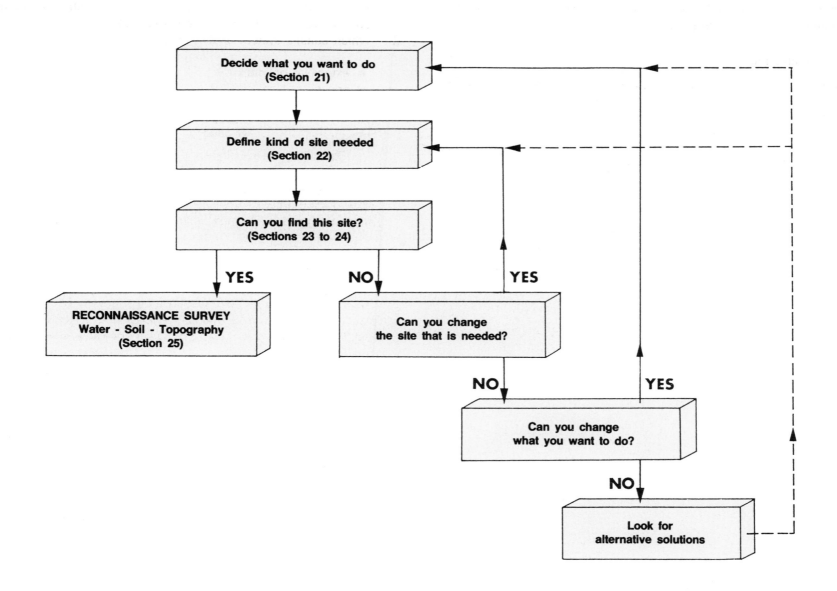

22 Major considerations

1. The major factors to be considered when selecting a site for the construction of a freshwater fish farm are **water supply, soil quality and local topography**. However, other factors also have some importance (see Section 23).

2. When you consider all the factors for selecting a site for a particular purpose, you may find that there are no suitable sites in your area. This can often happen; you then have to find out whether you can change your plans or ideas to meet the characteristics of the sites that are available. You may also be able to get assistance with this.

3. It is also important to remember that there is no point in carrying out a detailed survey or assessment of a site unless you are reasonably sure it meets the basic needs of the farming you plan to do.

Water supply

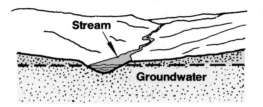

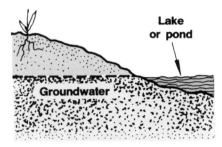

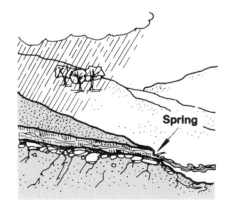

4. It is essential that you have the required supply of good quality water at the time needed to operate your fish farm. Preferably, it should be available all year round.

5. You have learned in a previous manual in this series (**Water, 4**):

- how to estimate the quantity of water you will need to operate your fish farm (2 to 5 l/s/ha) (Chapter 2);
- how to provide for such a water supply, if necessary from a small reservoir to be built on your farm (Chapter 4).

Note: you should always give preference to a site where you will be able to obtain your water supply by gravity*. If you do have to pump, try to minimize the distance, particularly the height (head) at which you need to pump (see also Section 39).

6. It is important that your water supply is of **good quality**. Try to get a supply which is reasonably clear of leaves, branches, plastic bags and other rubbish, as they can easily block your feeder canals or pond inlets. You should be particularly vigilant of the presence of chemical pollution. It may originate not only from certain industries but also from agricultural land where crops such as cotton, citrus and tomatoes are intensively produced. Pesticides can also be regularly applied to rice fields and to irrigation canals to control insects, snails and weeds.

Estimate quantity of water needed

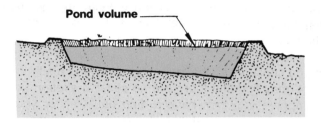

Pond volume

Estimate water losses

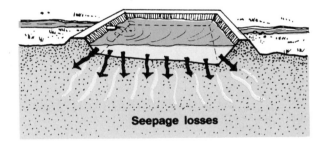

Seepage losses

Evaporation losses

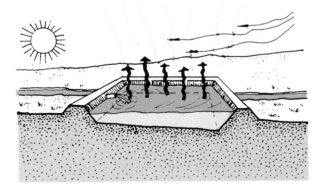

67

7. Find out also, if you can, whether there are any developments being planned, such as setting up a new factory, a new crop or crop technology, ploughing, quarry or construction works, which could affect the water quality.

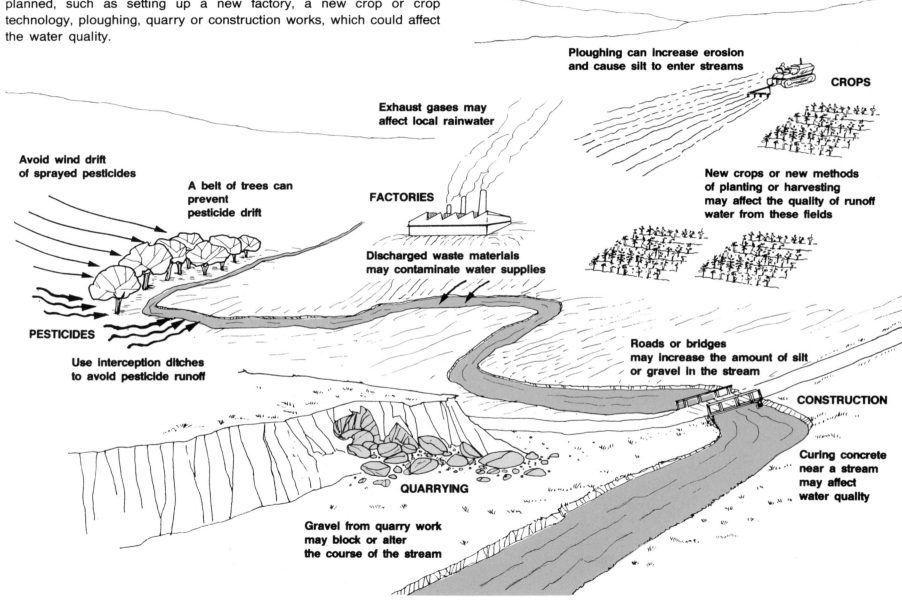

Ploughing can increase erosion and cause silt to enter streams

CROPS

Exhaust gases may affect local rainwater

Avoid wind drift of sprayed pesticides

A belt of trees can prevent pesticide drift

FACTORIES

New crops or new methods of planting or harvesting may affect the quality of runoff water from these fields

PESTICIDES

Discharged waste materials may contaminate water supplies

Use interception ditches to avoid pesticide runoff

Roads or bridges may increase the amount of silt or gravel in the stream

CONSTRUCTION

QUARRYING

Curing concrete near a stream may affect water quality

Gravel from quarry work may block or alter the course of the stream

Soil quality and site selection

8. You have learned in a previous manual in this series (**Soil, 6**) all you should know about soil to be used for fish farming. Remember to:

- avoid sites with rock outcrops, gravel beds, sandstone and limestone;
- beware of sandy soils and termite mounds;
- give preference to soils such as sandy clay, silty clay loam and clayey loam.

9. Remember that even a small layer of a troublesome material such as gravel, sand or acidic clay can cause problems. If there are such layers, make sure that the earthworks on the site do not enter into them.

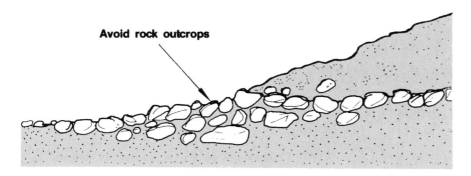

Avoid rock outcrops

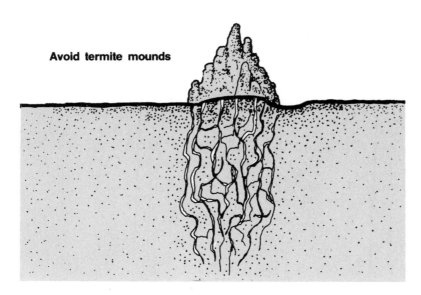

Avoid termite mounds

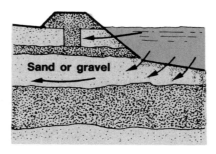

Sand or gravel

Water losses

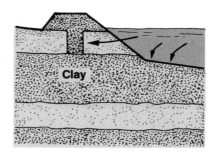

Clay

No water losses

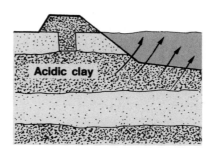

Acidic clay

Acidic clay may contaminate the water

Local topography and site selection

10. Local topography largely determines which type of pond you will build (see Section 16). The choice is based on the study of the **longitudinal profile** and **cross-sectional profiles** of the valley. You have learned in a previous manual in this series (Sections 95 and 96, **Topography**, **16/2**) how to do these studies either from existing maps or from your own measurements.

11. Look for sites:

- where **water drainage** will be possible by gravity*;
- where the **earthwork** will be minimum;
- where it will be easy to **balance the volume of earth** to be excavated and that to be filled in.

12. You will find such sites on gently sloping ground, where the slope is 0.5 to 3 percent. Avoid slopes greater than 5 percent. If you have to use horizontal land, it will be more costly to build drainable ponds.

Selecting a site

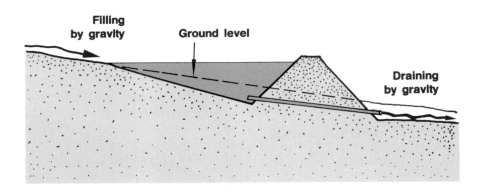

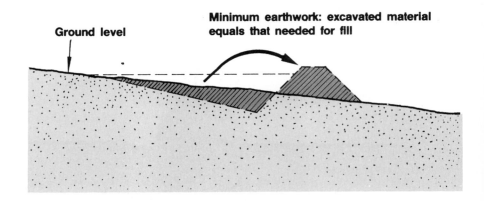

13. You will also have to ensure that the size of the selected site is large enough for your immediate purpose and, if necessary, for future expansion. A regular shape will make it easier to plan.

14. If you select a low-lying site, at the bottom of a valley for example, you should ascertain that the pond area will be free from deep flooding. Carefully observe the marks left by flood waters on bushes, trees, bridges, rocks or other permanent structures. Ask local people and specialized authorities, if necessary, to inform you about the expected depths of floods, in particular exceptional floods.

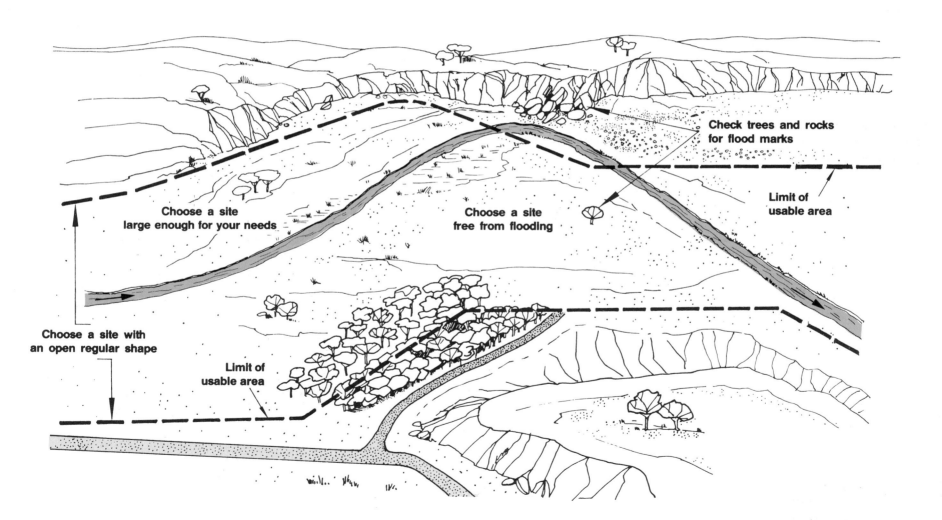

Choose a site large enough for your needs

Choose a site free from flooding

Check trees and rocks for flood marks

Limit of usable area

Choose a site with an open regular shape

Limit of usable area

23 Other important characteristics of the site

1. There are other site characteristics which are also important to consider during selection.

(a) **Vegetation cover:** if there are big trees or a dense population of smaller trees, clearing the land will be difficult and costly (see Section 52). Open woodland, grassland, old paddy fields or land covered with low shrubs permit easier and cheaper construction.

(b) **Accessibility:** the use of artificial feeds on a commercial farm, pond management and marketing will require good access by road to the site. On the other hand, for subsistence ponds or small-scale operations, access on foot or by bicycle or motorbike might be sufficient.

(c) **Proximity to your home:** it is always best to live close to your ponds. It will be easier to manage them and to protect them against poaching. For larger farms, a place must often be built for people who are looking after the ponds (see for example Section 18).

(d) **Multiple uses of the ponds:** it is sometimes advantageous to be able to use your ponds for purposes other than fish farming such as livestock watering, gardening or domestic use. Such integrated fish farming should be well planned (see Section 17, paragraphs 9 and 10).

(e) **Proximity and size of market:** once harvested, your fish should preferably be sold fresh, as soon as possible and with the least costs. You should know in advance how many fish you will be able to sell in one day and plan your pond sizes and harvests accordingly.

(f) **Availability of inputs:** if your fish farm requires regular inputs such as feed ingredients and juvenile fish, they should be available locally. You might also want to hire casual labour to help you from time to time. Spare parts and supplies might also be necessary.

Note: remember to ensure that there are **no legal restrictions** on the utilization of the site and the usage of water.

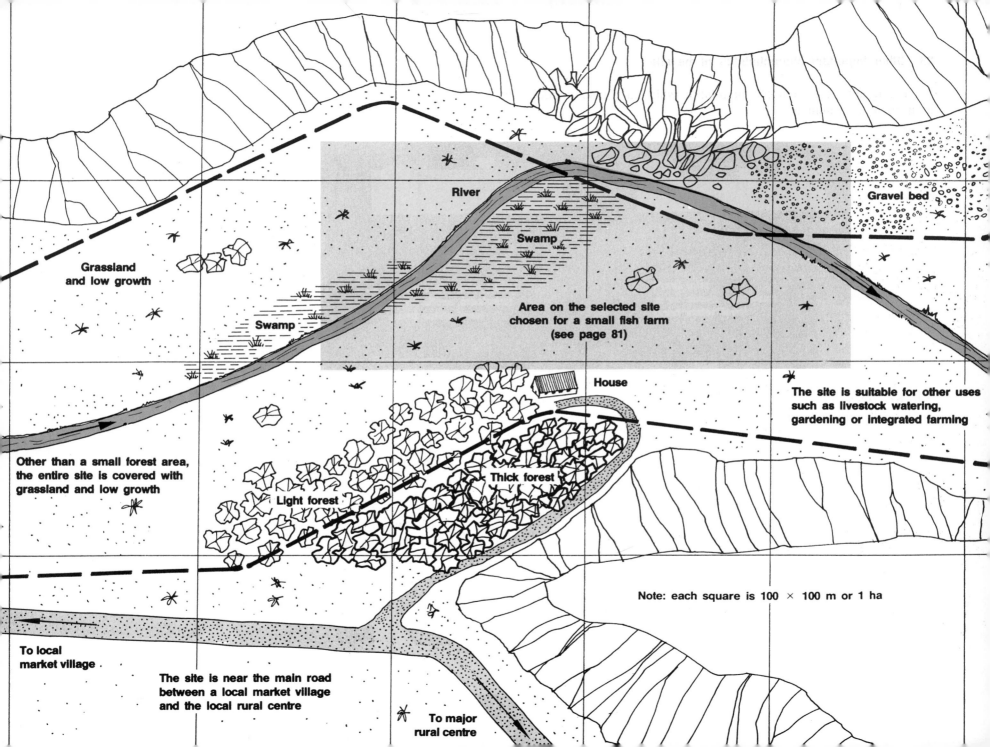

Grassland
and low growth

Swamp

River

Swamp

Gravel bed

Area on the selected site
chosen for a small fish farm
(see page 81)

Other than a small forest area,
the entire site is covered with
grassland and low growth

House

The site is suitable for other uses
such as livestock watering,
gardening or integrated farming

Thick forest

Light forest

Note: each square is 100 × 100 m or 1 ha

To local
market village

The site is near the main road
between a local market village
and the local rural centre

To major
rural centre

1.   Unless you know the region well, it will be difficult for you to locate quickly those sites whose features are likely to be favourable.

2.   Your field surveys will be easier if you can make **preliminary studies** from **existing topographical maps**, at the 1:50 000 scale for example (see Section 111, **Topography**, **16/2**). From the types of valley in the region and their longitudinal profiles, you can already decide on the possible types of pond to be built (see Section 16).

3.   If you decide on a **barrage pond**, you can estimate its drainage area and the water available through runoff*.

4.   The availability of **access roads** can be determined. If roads exist but are too far away, the potential of the site might be limited by the cost of building an access road.

5.   If available, other kinds of maps, such as soil maps and hydrological maps, can also be very useful.

TOPOGRAPHICAL MAP

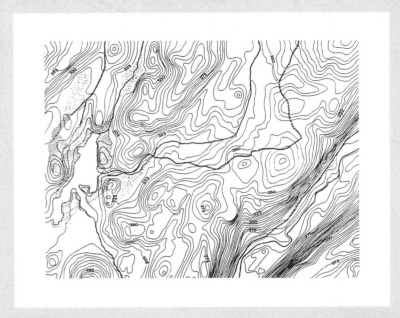

Scale 1:50 000

Note: an example of a topographical map at the scale 1:50 000 is shown on this page. On the opposite page this map is greatly enlarged to help illustrate how to use it. Note that various kinds of terrain are labelled as well as the location of a number of possible pond sites (see site A, B, C, D, E and F).

25 How to evaluate a potential site

1. Each potential site should be evaluated in the field through a series of quick **reconnaissance surveys** to ensure that the major requisites are met:

- availability of water supply (see **Water**, **4**);
- quality of water supply (see Chapter 2, **Management**, **21/1**);
- adequacy of soil quality (see Section 24, **Soil**, **6**);
- suitability of topography (see **Topography**, **16/1** and **16/2**).

2. How thorough you need to be with each of these surveys will depend on the size of the fish farm you plan to build.

3. For a small fish farm with an area less than 3 000 m², a small series of rough measurements will be sufficient.

4. For a larger fish farm where layout and gradients are critical, you will need more accurate measurements.

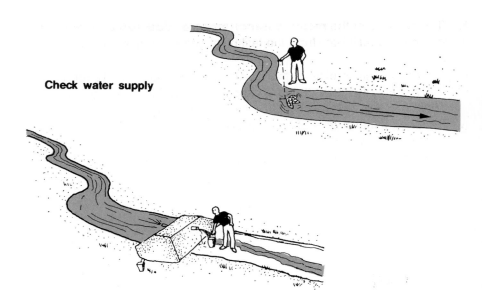

Check water supply

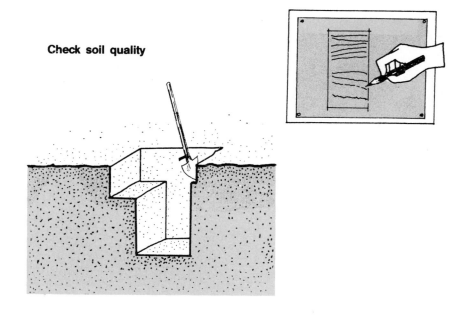

Check soil quality

5. The purpose of the **reconnaissance topographical survey** is to check if ponds can be built both from the technical and economic points of view.

(a) A longitudinal profile of the stream (see Section 112, **Topography, 16/2**) will give you the difference in elevation between various points. It will tell you if ponds can be built (see Section 18).

(b) A longitudinal profile of the site (see Section 82, **Topography, 16/2**) will confirm whether the site is also suitable (i.e. generally sloping) and would match well with the stream elevations.

Potential site

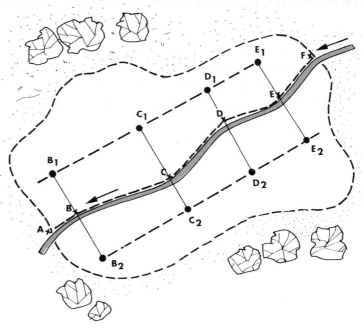

Longitudinal profile

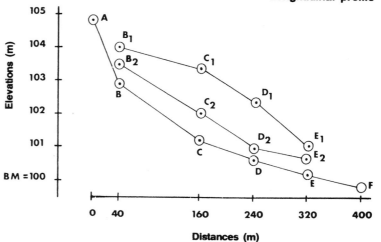

(c) Cross-section profiles of the site, perpendicular to the longitudinal profile, at 20- to 50-m intervals or less for small ponds and in irregular terrain, will give you the shape and size of the site.

6. From these measurements and other observations (see Sections 21 to 23), you should be able to **make a first evaluation of the site**. This evaluation should lead you either to reject or retain it as a potential site. In the latter case, you will need to study the site in more detail (see Section 26).

Potential site

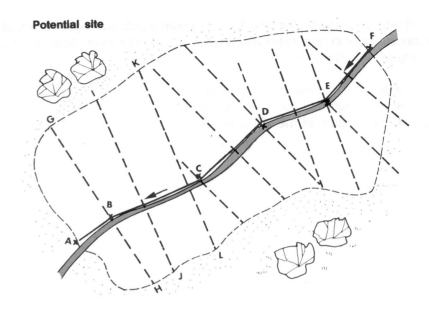

Cross-section profile GBH

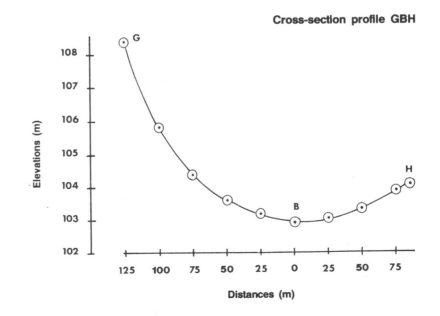

7. If you plan to use the site for a **barrage pond**, you could proceed immediately with the **feasibility study** of the site (see Section 113, **Topography**, **16/2**) by estimating:

- the area of the barrage pond;
- the volume of the barrage pond;
- the volume of the earthen dam.

8. Then compare these last two values. If the volume of the pond is greater than ten to 15 times the volume of the dam, and if the longitudinal slope of the farm site is less than 2 to 3 percent, the topography of the site can be considered acceptable (see Section 113, **Topography**, **16/2**).

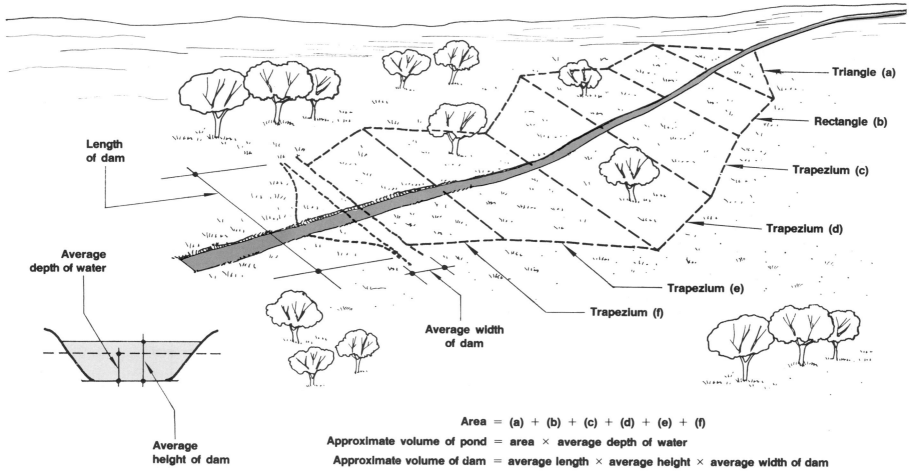

Length of dam

Average depth of water

Average width of dam

Average height of dam

Triangle (a)

Rectangle (b)

Trapezium (c)

Trapezium (d)

Trapezium (e)

Trapezium (f)

Area = (a) + (b) + (c) + (d) + (e) + (f)
Approximate volume of pond = area × average depth of water
Approximate volume of dam = average length × average height × average width of dam

9. However, when selecting a site for a barrage pond, remember that you should try to:

- build the shortest dam possible;
- make sure there is good material around the place where you want to build the dam;
- carefully check for possible floods; the smaller the drainage area, the smaller the floods (see Section 41, **Water**, **4**);
- if **springs** are your main source of water, avoid drowning them by having too great a water depth above them. You can sometimes test this by building a small earth wall around the spring to see how high it can push up the water. If this level is less than the intended water level in the pond, you will drown the spring when you build the pond. You can also test through a pipe how much water will flow at the intended pond depth.

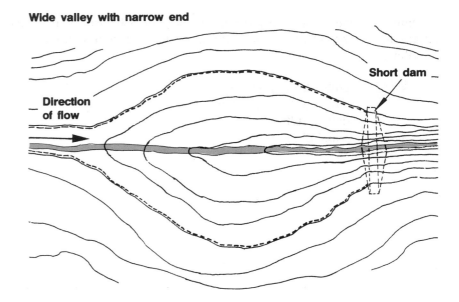

Wide valley with narrow end

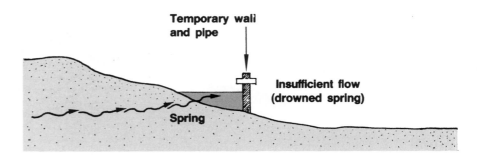

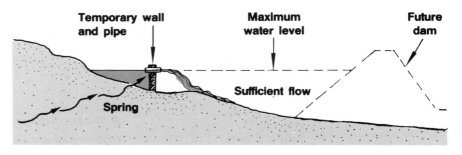

26 How to begin planning your fish farm construction

1. Now that you have selected the site of your fish farm, you should initiate the planning of the fish farm, which has two main related components:

- **the organizational planning**, in which you decide where, how and in which order you will build the farm (see Chapter 12, **Pond construction, 20/2**);
- **the physical planning**, in which you decide on layouts, detailed design and earthwork.

2. Below is a general layout of a small fish farm (see grey shaded area on page 73) chosen on the basis of soil and topographical surveys of the selected site.

3. The physical planning, which should be done in several steps, is discussed on page 83.

4. You will learn how to develop detailed plans and drawings for such a fish farm in Section 124, **Pond construction, 20/2**.

**Detail of the area on the selected site
chosen for a small fish farm (see page 73)**

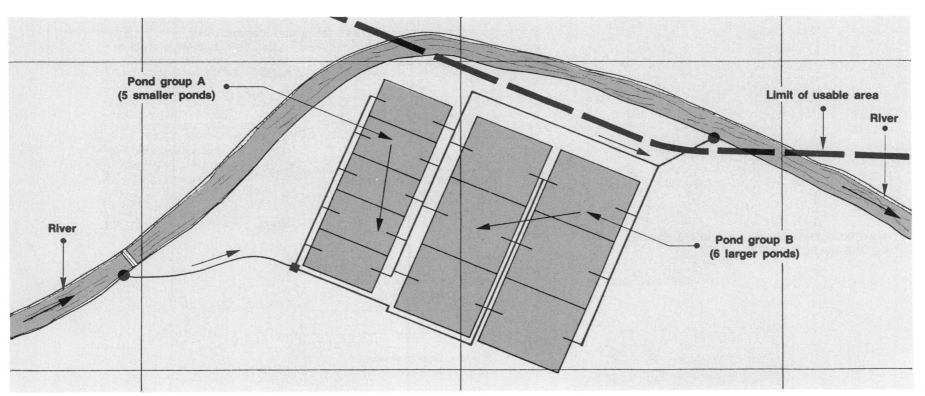

Pond group A
(5 smaller ponds)

Limit of usable area

River

River

Pond group B
(6 larger ponds)

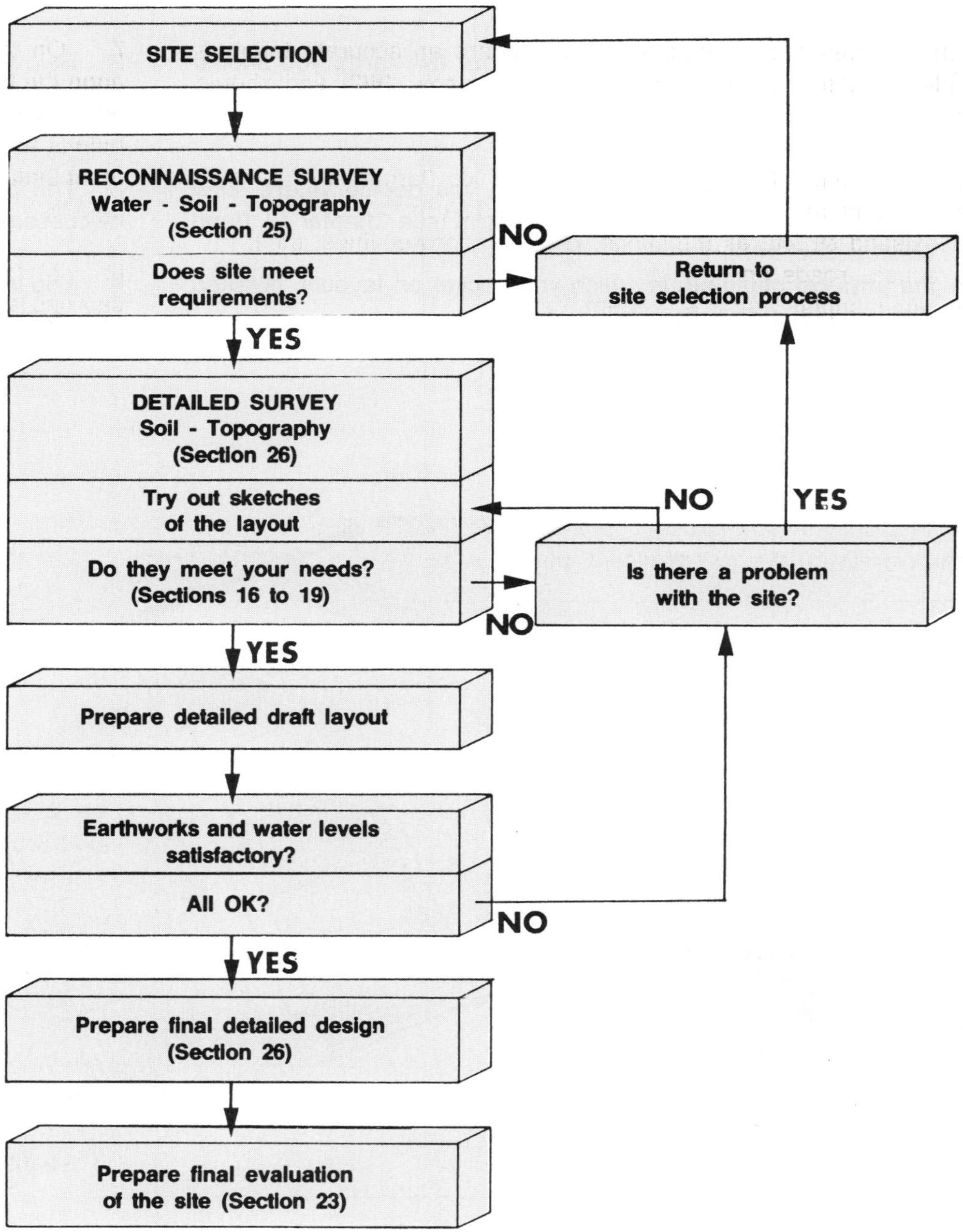

MATCHING THE FISH FARM AND ITS LAYOUT TO THE SELECTED SITE

SITE SELECTION

RECONNAISSANCE SURVEY
Water - Soil - Topography
(Section 25)

Does site meet requirements?

NO → Return to site selection process

YES

DETAILED SURVEY
Soil - Topography
(Section 26)

Try out sketches of the layout

Do they meet your needs?
(Sections 16 to 19)

NO ← Is there a problem with the site? → **YES** → Return to site selection process

NO

YES

Prepare detailed draft layout

Earthworks and water levels satisfactory?

All OK?

NO

YES

Prepare final detailed design
(Section 26)

Prepare final evaluation of the site (Section 23)

Detailed topographical survey of the site

5. Such a survey should enable you to prepare an accurate topographical plan of the site (see Chapter 9, **Topography**, **16/2**) and should include:

- scale from 1:1 000 (small farm) to 1:2 000 (larger farm);
- contours at a vertical interval of 0.20-0.50 m;
- existing structures (buildings, reservoir, power lines, etc.);
- access roads and tracks;
- water supply (springs, stream, canal);
- property boundaries, if close enough to site;
- particular features (rocks, pits, ant hills, trees);
- location of bench-marks and soil sampling stations.

Detailed soil survey of the site

6. Make a detailed soil survey of the site (see Section 25, **Soil**, **6**) and show the results on the topographical plan.

Layout of the fish farm

7. On the basis of all the information now available to you, decide upon the **type of pond** (or ponds) it is best to build (see Section 16), and which **layout** to adopt (see Sections 17 and 18). To do this, it may be helpful to make one or more simple versions of the site plan, with the main features identified, on which you can sketch out in pencil alternative layouts.

8. When you have worked out a layout which suits your needs, you can use a copy of the topographical plan of the site to prepare a more detailed **topographical plan of the fish farm** showing the number of ponds, their size and orientation, the dikes, the water supply system, the drainage system and other structures as necessary. You will learn more about all these in the following chapters.

9. You will have to check your layout carefully to see that:

- you dig out about the same **quantity of earth** as you need to build up (see Chapter 6);
- **water levels** are correct in the site to allow the ponds to fill and drain (see later chapters).

10. If these two points do not work out properly, you may have to amend the layout and then recheck earth and water levels.

11. Once you are satisfied that earthwork volumes and water levels are reasonably good and that the layout is suitable, you can proceed with the more detailed planning.

12. Begin by learning more about materials, equipment, various structures commonly used on fish farms, their function and how to build them. These will be described in the following sections of this manual.

13. You will then be able to plan the construction of your fish farm properly in more detail and according to your own requirements. This will be described in the second part of this manual (see **Pond construction**, **20/2**).

3   BASIC MATERIALS FOR CONSTRUCTION

1.   In addition to the pond soil itself, you might use a range of different materials for the fish farm, for example for foundations, water supply and water control devices. The choice ranges from locally available materials such as bamboo and wood to bricks, cement blocks, concrete and plastics (for pipes), for which you may have to go to specialized dealers.

Selecting materials

2.   The choice of construction materials essentially depends on their **suitability**, their **local availability** and the amount of **money** you are prepared to invest.

3.   If you are a beginner fish farmer and your farm is very small, it is best to use simple structures and not to spend too much on materials. As you gain experience and wish to expand, your investments may increase, and more permanent, better structures may be built.

4.   If you plan to build a large fish farm, you should select the most suitable permanent structures from the start.

Weight per unit volume of materials

5.   The materials generally have to be transported to the construction site. To plan this properly and help estimate the cost of transport and handling, use **Table 5**, which gives the weight per unit volume (kg/m^3) of common basic materials.

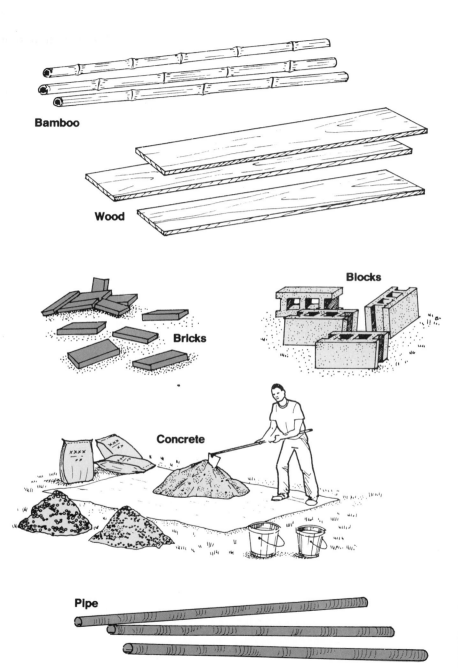

Bamboo

Wood

Blocks

Bricks

Concrete

Pipe

TABLE 5

Specific mass of various construction materials

Material	Specific mass (kg/m³)
Bamboo	300-500
Wood	500-1 100
Bricks	1 500-1 800
Cement blocks	1 500-2 000
Crushed bricks (for foundation)	950-1 250
Building stone, dry loose	1 400-1 600
Earth, dry loose	1 300-1 500
Earth, moist rammed	1 750-1 850
Gravel	1 300-1 500
Sand, dry to wet	1 450-2 000
Cement	1 250-1 400
Clay, dry compacted	1 400-1 500
Cement concrete	2 100-2 400
Cement mortar	2 000-2 200
Reinforced concrete (5% steel)	2 600-2 700

Note: for granular materials, these figures are bulk densities, i.e. including the pore space between the particles. The actual density is higher

Reinforcement steel bars
(see also Chart on page 135)

Diameter (mm)	Weight (kg/m)
6	0.222
8	0.395
10	0.617
12	0.888

31 Bamboo and wood

Special characteristics of bamboo

1.   Bamboo is a wooden perennial grass that lives for an indefinite number of years. It grows fast, occurring naturally throughout the world, but particularly in tropical Asia. The Chinese bamboo or "yellow bamboo" (*Bambusa vulgaris*) has been introduced in several African and Latin American countries where it is now found widely at altitudes between sea level and 1 500 m.

2.   Lengths of bamboo wood, or culms, are cylindrical and divided at intervals by raised nodes from which the branches grow. At each node there is a partition wall that completely separates the cavity of one internode from the next. Culms are covered outside and inside by hard waxy cuticles that offer considerable resistance to the absorption of water, particularly when properly dried out. Bamboo is strongest at the age of three to four years. Because of its versatile characteristics, bamboo has numerous uses, for example as a construction material, for water pipes and for erosion control.

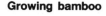

Growing bamboo

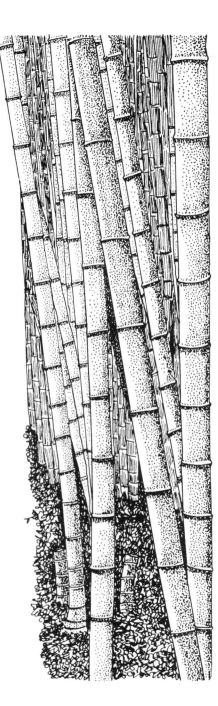

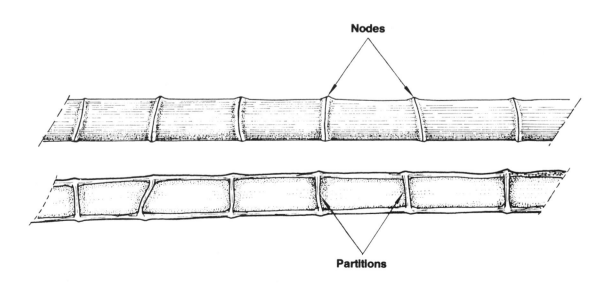

Nodes

Partitions

Producing good bamboo pipes

3.   When harvesting fresh bamboo, avoid letting them dry for too long; they shrink as they dry, and small cracks develop, which can weaken the bamboo for later use as a water pipe.

4.   To produce good quality bamboo pipes:

(a)   Cut mature bamboo and carry them out of the forest.
(b)   If necessary store bamboo in the shade, covering them with branches or large leaves.
(c)   Partially break or drill the partition walls inside the culms (see paragraphs 6 and 7).
(d)   As soon as possible, immerse the freshly cut bamboo into water (in a reservoir, river or pond).
(e)   To desap them, leave them in the water for six to eight weeks to extract the chemical substances present in the bamboo culm wall and to improve the durability of the pipes.
(f)   After desapping, remove all remaining parts of the partition walls in the bamboo pipe.
(g)   The bamboo pipe is ready for use.

Note: if you harvest bamboo during the dry season or at the beginning of the rainy season, desapping will be easier, and the quality of the pipes will be better.

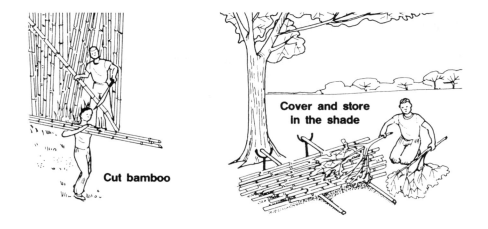

Cut bamboo

Cover and store in the shade

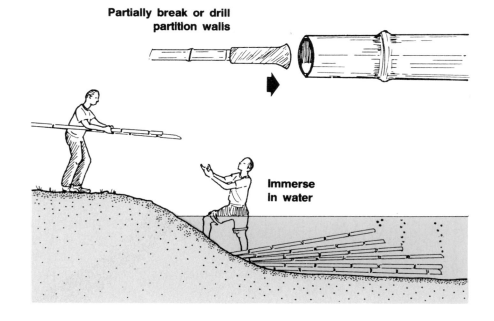

Partially break or drill partition walls

Immerse in water

Remove remaining partition walls

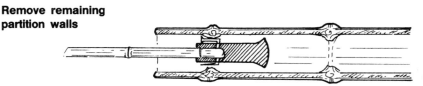

Removing the partition walls of bamboo

5. There are two simple ways of removing the partition walls of bamboo: drilling them or cutting them out. If the walls are tough, they might be too hard to drill.

6. Drill the partition walls manually with a circular bit, which you can easily make yourself.

(a) **Bell out** one end of a short length of steel pipe to increase the diameter.

(b) **Sharpen** the edge with a file.

(c) **Push** a length of bamboo of sufficiently small diameter into the pipe to act as a **handle**.

(d) Secure this bamboo to the pipe by **drilling a small hole** through the assembly and **inserting a nail** in the hole.

(e) **Bend the nail** to fix it into place, making sure it does not stick out too far and cause the drill to jam.

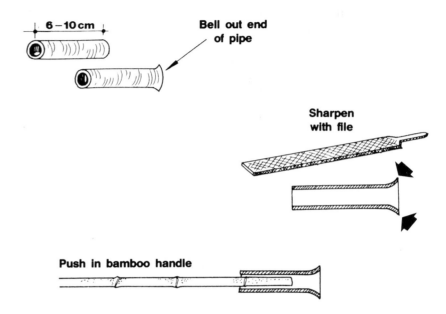

6 – 10 cm

Bell out end
of pipe

Sharpen
with file

Push in bamboo handle

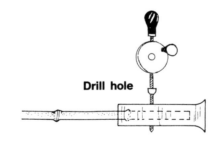

Drill hole

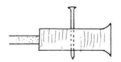

Insert and bend nail

(f) For each bamboo, **first use the smallest diameter bit** and bore a hole through each partition wall.

(g) Later, for example after desapping the pipes, **progressively enlarge** these holes with larger diameter bits.

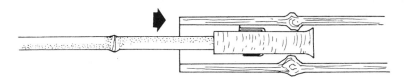

Push bit through partition walls

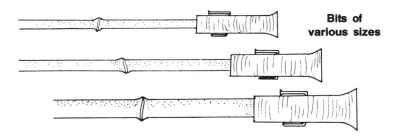

Bits of various sizes

Note: you may need several bits for various sizes of bamboo. To drill through or break partition walls, you may need the help of other people.

You may need one or more people to use the bit

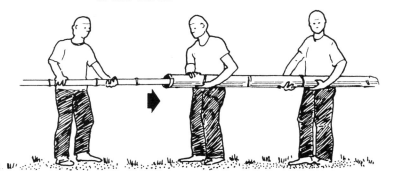

7. To cut out the partition walls, proceed as follows.

(a) **Fix the bamboo culm on the ground**, for example between strong stakes, to keep it from rotating.

(b) Using a saw, **cut a thin slit** on each side of the first node where the partition wall is known to be.

(c) Using a sharp wood chisel, **cut out a small square piece** from the top part of the bamboo culm. Cut it as cleanly as possible.

(d) **Keep all the squares** that you cut out. You will need them later.

Fix bamboo with strong stakes

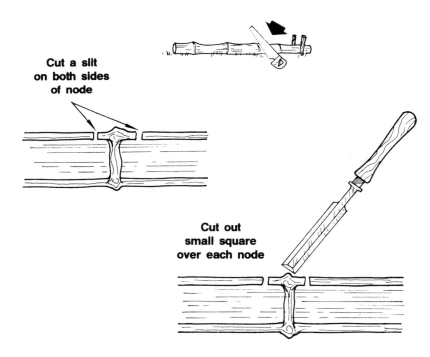

Cut a slit on both sides of node

Cut out small square over each node

Keep the squares

92

(e) Through the hole, **cut out the partition wall** with the wood chisel.

(f) Proceed as above for each partition wall, making sure that the square hole is always cut along the same line on top of the bamboo.

(g) When you have removed the last partition wall, place the bamboo vertically and take out all the loose pieces from inside the bamboo.

(h) Put the little square pieces you have kept aside back into the holes and secure them with a piece of string or wire.

(i) Cut one end of the bamboo obliquely. The water pipe is now ready.

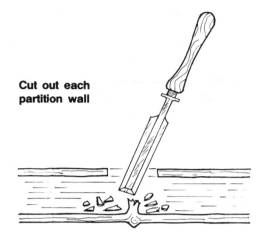

Cut out each partition wall

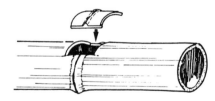

Put back each square

Tie or wire square in the hole

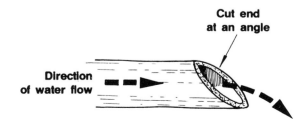

Cut end at an angle

Direction of water flow

93

8.   Bamboo used for construction is cut, immersed and stored in the same way as the bamboo for pipes, although cutting internal walls is not so important. However, if you are using the bamboo for piling (driving them into the ground), removing the walls helps anchor the bamboo.

9.   Wherever possible use lashings or ties to connect the pieces, as nails and screws will cause splitting and will weaken the structure.

Remove partitions when driving bamboo into the ground

Earth inside helps to anchor the bamboo

Broken partitions

Note: see use of bamboo as a stream barrier on page 15, Pond construction, 20/2

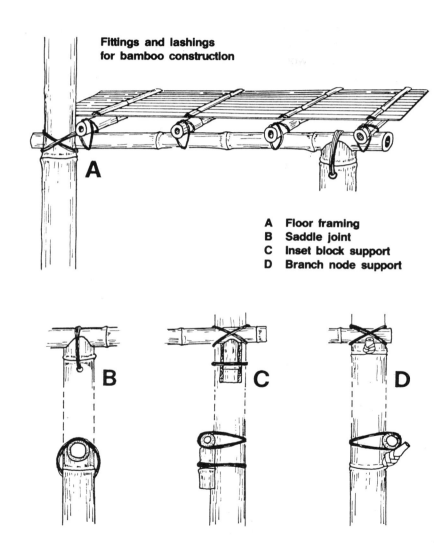

Fittings and lashings for bamboo construction

A

A   Floor framing
B   Saddle joint
C   Inset block support
D   Branch node support

B   **C**   **D**

Diverse characteristics of wood

10. The characteristics of wood, in particular **density***, **hardness** and **natural durability**, vary greatly. It is best to select the wood variety according to its use (see **Table 6**):

- **very durable wood** can be permanently exposed to humidity and kept in contact with the soil. It is usually very resistant to rot, termites and woodborers;
- **durable wood** resists air humidity well but should not be in permanent contact with the soil without treatment for preservation;
- **non-durable wood** should preferably not be used under humid conditions or in contact with soil. Preservatives can sometimes be used to treat the wood, but they do not usually provide permanent protection.

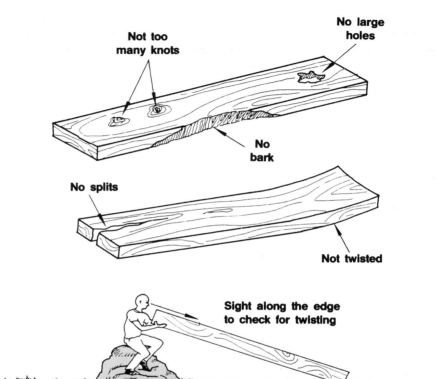

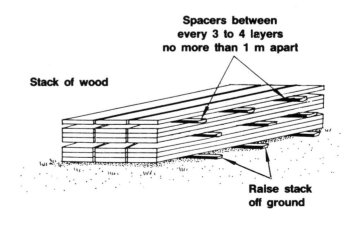

TABLE 6

Characteristics of selected wood varieties

Wood variety: common names (Latin genus name)	Density	Hardness	Natural durability*
Ako, antiaris, bonkonko *(Antiaris)*			1
Okoume *(Aucoumea)*	<0.50	0.2-1.5	1
Fromager, fuma *(Ceiba)*	very light	very soft	1
Samba, obeche, wawa, ayous *(Triplochiton)*			1
Abura, bahia *(Mitragyna)*			1
Ilomba, lolako *(Pycnanthus)*	0.50-0.64	1.5-3	1
Limba, afara *(Terminalia)*	light	soft	1
Framire, idigbo *(Terminalia)*			1-2
Acajou *(Khaya)*			2
Sapelli, aboudikro *(Entandrophragma)*			2
Sipo, utile *(Entandrophragma)*			2
Teck *(Tectona)*			2
Bete, mansonia *(Mansonia)*	0.65-0.79	3-6	3
Bilinga, opepe *(Nauclea)*	semi-heavy	semi-hard	3
Iroko *(Chlorophora)*			3
Makore, douka *(Tieghemella)*			3
Mukulungu *(Autranella)*			3
Moabi *(Baillonella)*			3
Doussie, afzelia *(Afzelia)*	0.80-0.95	6-9	3
Tali *(Erythrophloeum)*	heavy	hard	2
Niové *(Staudtia)*			2-3
Azobé, ekki, bongossi *(Lophira)*	≥0.95 very heavy	9-20 very hard	3

* 1 = non-durable; 2 = durable; 3 = very durable

11. Wood used as a permanent construction material should be free from bark wood and should not have any large holes or too many knots. It should be well dried and should not be twisted or split. It should be stored flat in a dry place with good air circulation.

12. For temporary use, e.g. to make casting forms for concrete (see Section 34), use light, cheap wood. If you are re-using it, make sure that the surfaces facing the concrete are even and are free from nails or splinters. To build water control structures, use heavier, preferably very durable wood, such as iroko or makore.

13. To increase durability, especially of wood in permanent contact with the soil, you can treat its surface.

(a) Burn the surface of the wood (e.g. the bottom part of poles).
(b) Use tar (e.g. on the bottom part of poles or the outside of a structure near soil). If you can, it is better to apply hot tar.
(c) Use waste motor oil diluted with a solvent such as paraffin which, when applied, will penetrate the wood and will help to force moisture out.

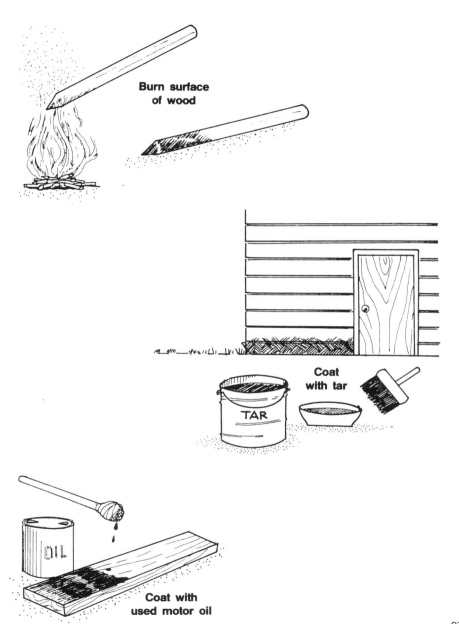

Burn surface of wood

Coat with tar

Coat with used motor oil

(d) Use special wood preservatives. These are more expensive and are usually copper, lead, zinc, or tin compounds in a solvent. As these are poisonous materials they should be handled very carefully.

Use special wood preservatives

Note: for the best results with tar, oil or other preservatives, apply the preservative generously and allow plenty of time to let it get into the wood. Put on several coats of preservative or dip the wood into a can or trough of preservative for at least 30 minutes. Make sure the end grain of the wood is properly treated, as this is where decay often starts.

If you use a brush, apply several coats

30 minutes

If you dip, leave for at least 30 minutes

Clay bricks

1. There are many different kinds of bricks. **Lightweight hollow bricks** are usually not strong enough for fish farm use. **Solid burnt clay bricks** are commonly used for fish farms. They are made of clay, air dried and fired in a special kiln. Their quality greatly depends on this firing. You should reject bricks which are too uneven, cracked, and either burnt too much or not enough.

2. In places where they are available, **industrially made bricks**, either solid or with a slight hollow on each side, or with two or three small "finger holes", are also suitable. **"Engineering bricks"**, usually a yellow/black colour, are also useful for foundations and heavily loaded areas, as they are much stronger and resist penetration by water.

3. Bricks are generally available in standard sizes, which vary from country to country. The following are examples of various standard sizes: **4 × 10.5 × 22 cm, 6 × 10.5 × 22 cm, 7.5 × 10 × 20 cm** and **10 × 10 × 20 cm**.

4. Bricks are used together with cement mortar (see Section 33). They should be stored and handled carefully to avoid excessive breakage. They should be well soaked in water for at least 30 minutes before use.

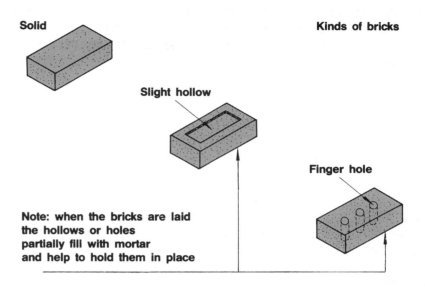

Kinds of bricks

Solid

Slight hollow

Finger hole

Note: when the bricks are laid the hollows or holes partially fill with mortar and help to hold them in place

Various standard sizes for bricks

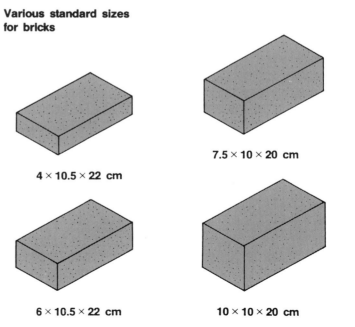

4 × 10.5 × 22 cm

7.5 × 10 × 20 cm

6 × 10.5 × 22 cm

10 × 10 × 20 cm

Cement or concrete blocks

5. Cement or concrete blocks are made of a mix which is cast and pressed into a special form. Concrete blocks can be made on site if needed, but care should be taken to make sure they are properly cast (see Section 34). Blocks should be at least 28 days old before they are used for construction.

6. Both hollow and solid cement blocks are produced. They are available in several standard sizes, usually with a length of from 40 to 50 cm, a height of 20 cm and a thickness of from 5 to 20 cm. The following are examples of various standard sizes: **5 × 20 × 40 cm**, **10 × 20 × 40 cm** and **20 × 20 × 40 cm**. Blocks are sometimes available in different weights: generally the heavier the block, the stronger. Blocks are used together with cement mortar (see Section 33). They should be properly stored and handled. They should be well wetted with water before use.

7. Standard clay bricks and cement blocks have a low resistance to humidity. They should, therefore, preferably not be used for foundations or underground constructions. In the presence of water, they should be well protected by impervious surfacing made of rich mortar (see Section 33).

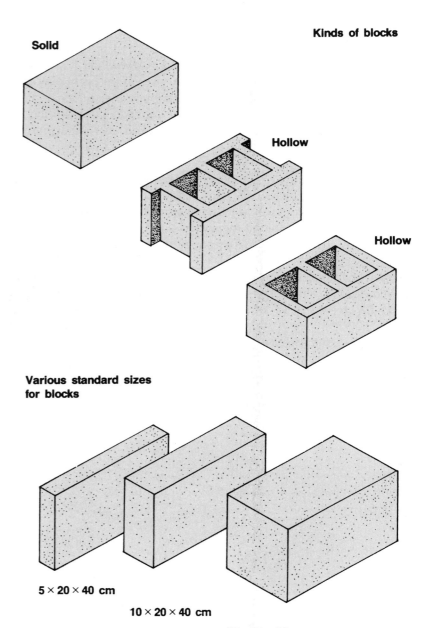

Kinds of blocks

Solid

Hollow

Hollow

Various standard sizes for blocks

5 × 20 × 40 cm

10 × 20 × 40 cm

20 × 20 × 40 cm

8.   Stones are used in some areas for construction, usually for walls and for lining water canals, dikes and spillways. Their characteristics depend on the type of rock they come from (see **Soil, 6**):

- **sedimentary:** sandstones, siltstones, shales, limestone. These are even, regular, often quite soft with rounded corners, and can often be split or cut to make regular-shaped blocks;
- **igneous:** granite, basalt, pitchstone, pumice. These have various properties and are often very hard and strong and of an irregular structure. It is difficult to make blocks from them. Soft lava rocks are very light and weak;
- **metamorphic:** marble, quartzite. These are often irregular, hard, very tough and difficult to shape. Slates can be split to make flat pieces and can be useful for stopping movement of water.

9.   Stones are used either **"dry"**, without any mortar or jointing material, by carefully selecting and fitting them together or, more commonly **"wet"**, by setting them in mortar.

10.   For walls, unless you have cut stones, it is usually important to have a range of sizes, using smaller stones to fill in the gaps and secure the larger stones in place.

11.   You will also need larger stones at the corners, at intervals along the wall and across the width of the wall, to give the wall strength and stability.

12.   Usually, stones with irregular, rough edges make stronger walls. For lining canals, small, smooth, rounded stones are best as they let water flow past more easily.

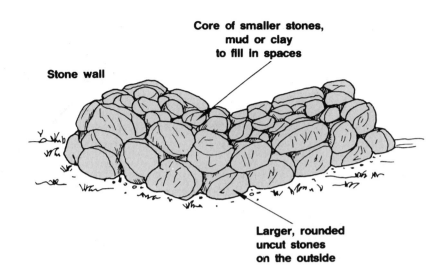

Stone wall

Core of smaller stones, mud or clay to fill in spaces

Larger, rounded uncut stones on the outside

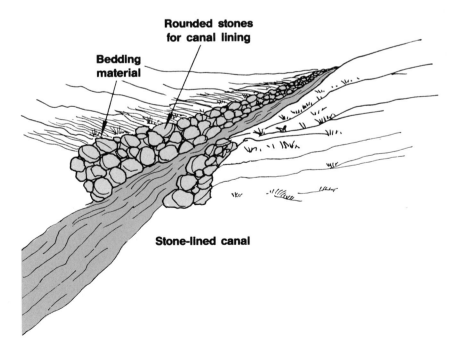

Bedding material

Rounded stones for canal lining

Stone-lined canal

33 Cement mortars

1. A cement mortar is a well-proportioned mixture of sand, cement and water. It is mainly used for joining together and surfacing materials such as stones, bricks and cement blocks. A good mortar is homogeneous, soft and shiny. It has a smooth appearance and a plastic consistency.

2. To prepare a good mortar, it is very important to use the proper ingredients and to mix them thoroughly in the right proportions.

Selecting the sand

3. It is best to use a clean, well-graded mixture of coarse to fine sand, with particle sizes varying from 0.2 to 5 mm. If possible, you should avoid using sea beach sand or sand deposits contaminated with salts. If you have to use these materials, you should wash the sand well.

4. Natural sand deposits can sometimes be found not too far from the construction site, for example in stream beds, dried-up lake or stream areas or quarries, but well-graded sands seldom occur naturally. In many cases, you will have to sift the sand through a 0.2-mm mesh screen to eliminate the finest particles. If there are particles larger than 5 mm, you should also eliminate them, using a 5-mm mesh screen.

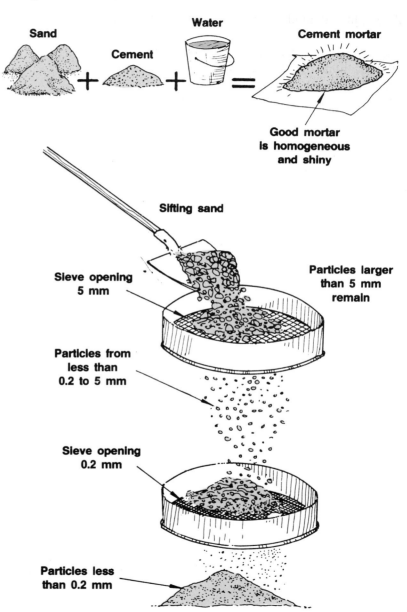

Mixing cement mortar

Sand + Cement + Water = Cement mortar

Good mortar is homogeneous and shiny

Sifting sand

Sieve opening 5 mm

Particles larger than 5 mm remain

Particles from less than 0.2 to 5 mm

Sieve opening 0.2 mm

Particles less than 0.2 mm

Checking the cleanliness of the sand

5. The sand should be free from silt, clay and organic materials.

6. A simple test to check the cleanliness of the sand is as follows:

(a) Obtain a clear, wide-mouthed glass jar.
(b) Fill the bottom of the jar with sand 5 cm deep.
(c) Add clean water until the jar is three-quarters full.
(d) Add, if available, two teaspoons of common table salt per litre of water.
(e) Close the jar and shake it vigorously for one minute.
(f) Let it stand for three hours.
(g) Check the sand surface. If there is silt present, it will form a layer on top of the sand.
(h) If there are more than 3 mm of silt, the sand must be washed.

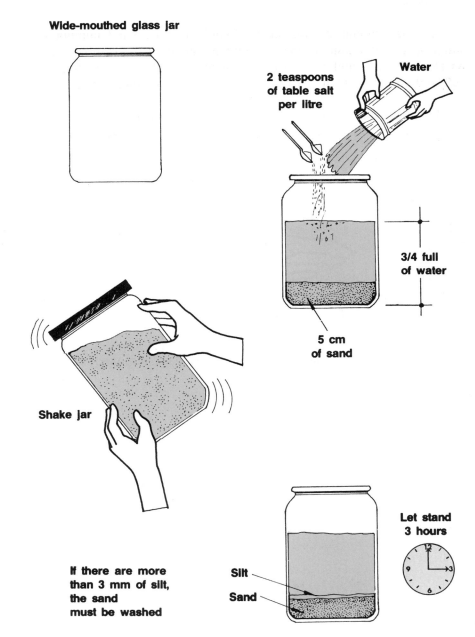

Wide-mouthed glass jar

Water

2 teaspoons of table salt per litre

3/4 full of water

5 cm of sand

Shake jar

If there are more than 3 mm of silt, the sand must be washed

Silt

Sand

Let stand 3 hours

7. Another simple test to check the cleanliness of sand is as follows:

(a) Take a handful of sand and squeeze it.

(b) Throw it away.

(c) If your hand is clean and free of sticky dust, the sand is clean.

(d) If your hand is dirty and sticky, the sand is dirty.

Washing the sand

8. If there is too much silt, you will have to wash the sand before using it. Repeat the following procedure until all your sand is clean.

(a) Put sand in a large, clean container such as a 200-litre metal drum.

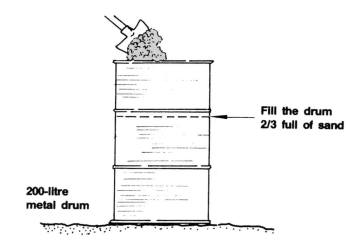

Fill the drum 2/3 full of sand

200-litre metal drum

(b) Cover the sand with clean, fresh water.

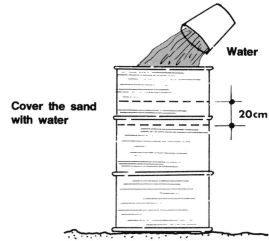

Water

Cover the sand with water

20 cm

(c) Stir vigorously.

(d) Let it stand for a few minutes.

**Stir vigorously
and let
stand**

(e) Pour off the dirty water.

(f) Repeat until the water remains clear.

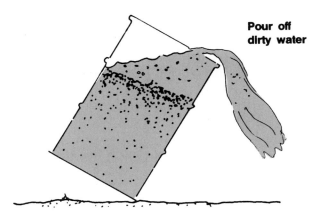

**Pour off
dirty water**

(g) Store the clean sand, so as to keep it from being recontaminated.

9.   You can store clean sand on the construction site, for example on a timber floor with low side walls as shown below.

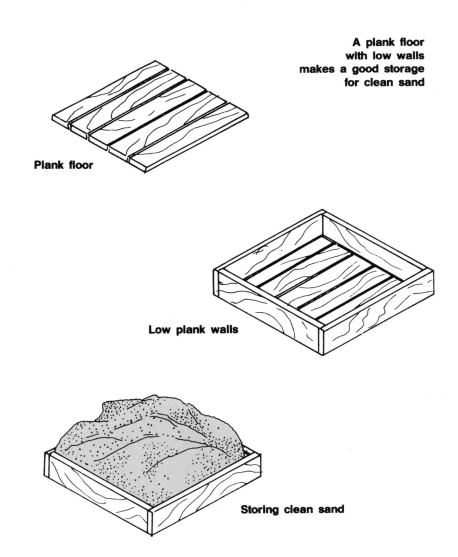

**A plank floor
with low walls
makes a good storage
for clean sand**

Plank floor

Low plank walls

Storing clean sand

Selecting the cement to use

10. You should use **ordinary Portland cement** (OPC), which is the standard, most widely available type of cement. It has the characteristics of setting and hardening in the presence of water while producing heat and shrinking in volume.

11. Remember that a mortar too rich in cement will crack while hardening.

Note: Portland cement strongly deteriorates in the presence of waters rich in calcium sulphate (more than 0.5 g/l) or sodium chloride (more than 4 g/l). In such cases, for example in acid sulphate soils or near brackish waters, a sulphate-resistant cement should preferably be used for constructions (see Section 18, **Soil**, **6**). This particular type of cement should never be mixed with Portland cement. If it is not available, use a slightly richer mix of OPC (see paragraph 19 in the next column), taking particular care in mixing, using and setting and making sure the cement has cured well before letting it come into contact with the soil or water.

12. Portland cement is usually classified according to its potential resistance to compression, which is usually either about 250 kg/cm^2 or about 325 kg/cm^2. For fish farm constructions you mostly use the "250" grade.

13. Portland cement is sold in thick paper sacks. The weight and volume of the sacks vary according to the country:

- **European system:** weight is 50 kg and volume about 40 l;
- **American system:** weight is 42.6 kg and volume about 28 l.

14. Check which system is used in your country to avoid errors when preparing cement mixes.

15. To guarantee the top quality of your cement, you should take the following precautions.

(a) Check the **freshness** of the cement before buying. It should be free from any lumps which cannot be pulverized between thumb and forefinger.

(b) Bring on site **only the number of sacks needed** for a short period.
(c) Protect your cement from **humidity**. Store it off the ground (a simple wooden platform is suitable) in a dry, well-sheltered place.
(d) Use the cement while it is **as fresh as possible**, and rotate your stocks properly.
(e) **Never** use hardened cement, but throw it away.

Selecting the water to use

16. The water should be clean and neutral or slightly alkaline (pH 7 to 8.5). It should be free from organic matter, oil, alkali or acid. Avoid using saltwater or a water too rich in sulphates (more than 250 ppm).

17. If you have to use brackish water or dirty water, add a tablespoon of soap powder for every sack of cement used. Dissolve the soap into a small amount of water and add it to the mix.

Selecting the mortar to prepare

18. There are three basic types of mortar which you can prepare yourself, as shown in **Table 7**, according to their use. Remember that the richer a mortar is in cement, the more it will shrink, and the more it will be inclined to crack.

19. In acid sulphate soils, the Portland cement proportion is normally increased by 10 to 20 percent.

20. If you need only small quantities of mortar, you can mix cement and sand by volume, in the following proportions:

- **lean mortar:** one part cement (250 grade) and four parts sand;
- **ordinary mortar:** one part cement (250 grade) and three parts sand;
- **rich mortar:** one part cement (preferably 325 grade) and two parts sand.

21. You will need about 200 l of water per m^3 of mix (about one part water for five parts mix).

Measuring the mortar components

22.   To produce a good mortar, it is essential that you measure accurately the amounts of cement and sand to be mixed, according to the proportions required.

23.   If you know the weight of one sack of cement, it is easy to calculate how many sacks you will need to use (see **Table 7**).

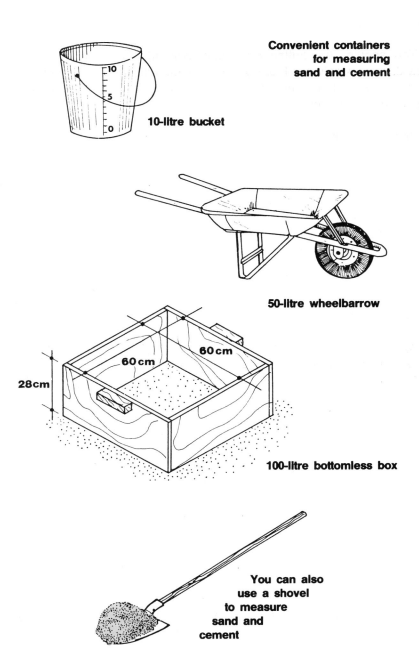

Convenient containers for measuring sand and cement

10-litre bucket

50-litre wheelbarrow

100-litre bottomless box

You can also use a shovel to measure sand and cement

TABLE 7

Basic types of cement mortars

Type of mortar	Portland cement		Sand* (l)	Utilization
	Quality	kg*		
Lean	250	350	1 000	Normal brickwork, pipe joints
Ordinary	250	450	1 000	Surfaces exposed to the air
Rich	325**	600	1 000	Waterproof surfacing, rejointing of brickwork

* Quantity of the material to prepare 1 m³ of mortar
  with the addition of about 200 l of water
** If available

24.   If you plan to use the above proportions by volume, it is best to use a container of known volume, such as a 10-litre bucket or a 50-litre wheelbarrow. For larger quantities, you can easily build yourself a 100-litre bottomles wooden box with handles as shown.

25.   A shovel may also be used for sand and cement, but you should take care to load the same amount each time. Even then this method is not very accurate.

Preparing a good mortar

26. To prepare a good mortar, carefully proceed as follows:

(a) **Prepare a clean mixing area**, for example a metal sheet or a watertight wooden platform. As a rough guide, a 1-m² area is enough for 50 kg of mix.

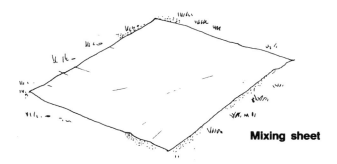

Mixing sheet

(b) **Measure** the quantity of sand required. If it is very dry, wet it a little before measuring.

Wet sand if dry

Measure sand

(c) **Spread the sand** over the mixing area.

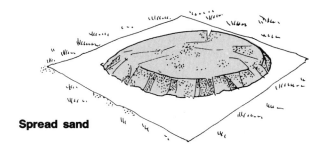

Spread sand

(d) **Measure** the quantity of cement required.
(e) **Spread the cement** on top of the sand.

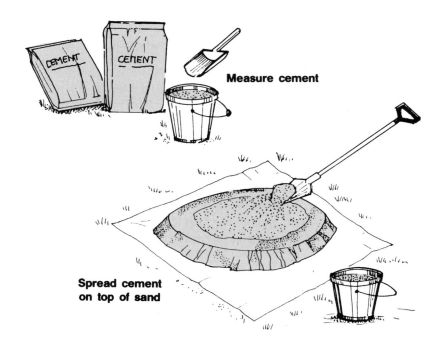

Measure cement

Spread cement on top of sand

(f) **Mix the sand and cement together thoroughly**, until the mix has a homogeneous colour. Be sure to mix in the bottom and side materials.

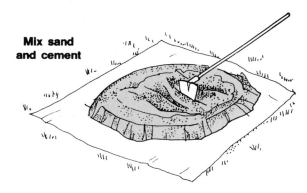

Mix sand and cement

(g) **Form a hollow** in the middle, slowly add a little water in the hollow and moisten part of the mix. Work the water in by carefully moving the dry mix in toward the hollow. Be careful not to let water run away.

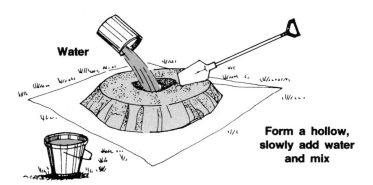

Water

Form a hollow, slowly add water and mix

(h) **Repeat adding water** little by little until the whole mix is moistened. Continue mixing thoroughly, adding just enough water to obtain a plastic consistency. The mortar should have a firm, smooth appearance. You should be able to make a clean slice into it with a trowel or shovel. It should sit on a trowel cleanly and firmly without loss of water and should spread smoothly.

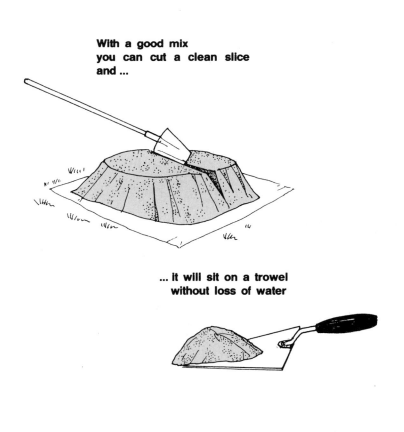

With a good mix you can cut a clean slice and ...

... it will sit on a trowel without loss of water

27. **Remember:** do not add too much water.

28. The mortar should be used immediately after its preparation. A mortar should never be used after it has started to set, which you can tell is happening if the mix starts to become stiff and breaks up when spread. Avoid using mortar which has dropped from the working area.

29. The surfaces to come in contact with the mortar should be clean and rough. It is essential to wet them well before applying the mortar, for example by soaking bricks in water for 30 minutes and wetting cement blocks, so that they do not absorb the water from the mortar and reduce its strength. If you are working in dry conditions, be careful to keep the bricks or blocks wet.

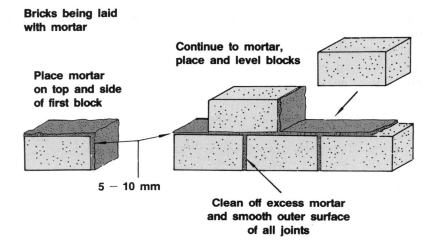

Bricks being laid with mortar

Place mortar on top and side of first block

Continue to mortar, place and level blocks

5 – 10 mm

Clean off excess mortar and smooth outer surface of all joints

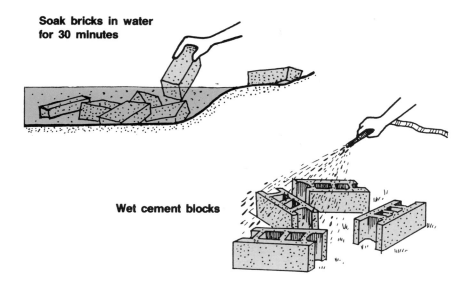

Soak bricks in water for 30 minutes

Wet cement blocks

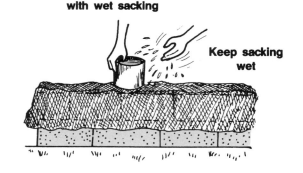

Cover fresh mortar with wet sacking

Keep sacking wet

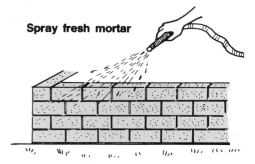

Spray fresh mortar

30. Protect mortar from the sun's heat and from drying wind until it hardens to the point where its surface cannot be scratched with a fingernail. At this stage, setting is complete enough for normal requirements. In hot, dry conditions, you can protect the setting mortar by covering the areas concerned with wet sacking, or alternatively, by using 110 a fine spray of water. However, be careful not to wash out the mortar.

34 Cement concrete

1. A cement concrete is a well-proportioned mixture of aggregates, cement and water. The **aggregates*** should be well graded so that when mixed they fit together with minimum pore space between them. These small remaining pores are filled up by the cement, which will then strongly bind the aggregates after its reaction with water.

2. The most important factors in making strong concrete are therefore:

- use of **properly graded aggregates*** with the right size and the right shape;
- addition of the **right amount of water**;
- **avoidance of very fine particles** as these will fill the small pores which should be filled with cement.

3. There are two to three different grades of **aggregates***, depending on the type of concrete required:

- **fine aggregates**, sand and rock screenings, size 0.2 to 5 mm. Sometimes termed "sharp" sands, these are usually coarser than sands used for mortar;
- **coarse aggregates**, gravel/pebbles, broken bricks, size 5 to 25 mm;
- **very coarse aggregates**, broken stones or bricks, size 25 to 60 mm.

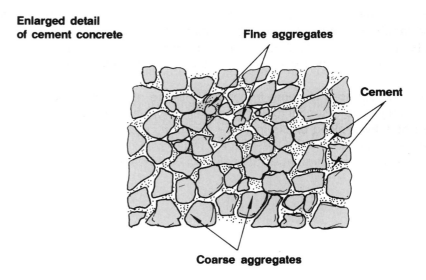

Enlarged detail of cement concrete

Fine aggregates

Cement

Coarse aggregates

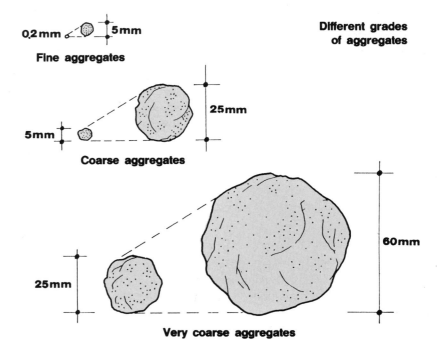

0.2 mm — 5 mm

Fine aggregates

Different grades of aggregates

5 mm — 25 mm

Coarse aggregates

25 mm — 60 mm

Very coarse aggregates

111

4. Good natural sources of aggregate materials, sand and gravel for construction purposes (**Table 8**) are relatively infrequent. Well-graded soils that have the right range of particle sizes are particularly hard to find. If the soils contain some silt, they are only moderately suitable.

5. Soils where silt and clay predominate are poor natural sources of aggregates. Unsuitable soils are all those belonging to the other Unified Soil Classification (USC) groups (see Section 111, **Soil**, **6**).

6. Good concrete is a homogeneous mix without excess water. It is smooth and plastic. It is neither too wet and runny, nor too dry and crumbling.

7. To prepare good concrete, you should use the proper ingredients and mix them thoroughly in the right proportions. You have already learned which kinds of sand, cement and water to use (see Section 33). In the following sections, you will now learn about coarse aggregates and concreting.

TABLE 8

Suitability of soils* as sources of aggregates for general construction**

Good sources	Moderate suitability	Poor suitability	Unsuitable soils
SW	SW-SM	SM	ML
SP	SP-SM	SW-SC	CL
GW	GW-GM	SP-SC	OL
GP	GP-GM	GM	MH
		GP-GC	CH
		GW-GC	OH
			Peat

W: well-graded materials; P: poorly graded materials; S: sand; G: gravel; C: clay; M: silt; O: organic; L: low or H: high plasticity/compressibility
*   Soils are defined according to the Unified System of Classification (USC), see Section 111, **Soil**, **6**
** These suitability classes apply for use in concretes and in dry aggregates (e.g. for mending roads, etc.)

Selecting the kind of gravel and broken materials to use

8.   Your concrete is only as strong as the strength of its coarse aggregates. Therefore you should look for hard, dense and durable gravel and stones. These aggregates should never be lateritic (see Section 18, **Soil**, **6**).

9.   If you have doubts about the strength of the coarse materials to be used, you can use the following test.

(a)   Break some stones with a hammer.
(b)   Break similar sizes of concrete pieces, and compare how difficult these are to break.
(c)   The stones are hard enough if they are more difficult to break than the concrete.

10.   These coarse aggregates should preferably be neither flat in shape nor have sharp edges. The best materials have **round or cubical shapes** such as gravel from a stream bed or beach.

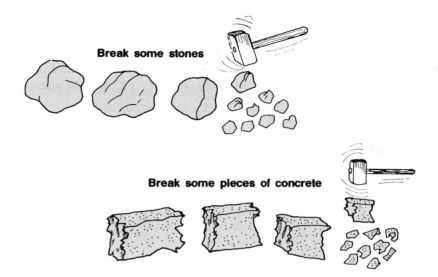

Break some stones

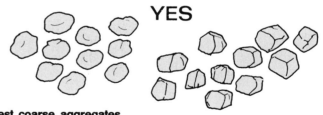

Break some pieces of concrete

The stones are hard enough to use if they are more difficult to break than the concrete

YES

The best coarse aggregates are round or cubic in shape ...

NO

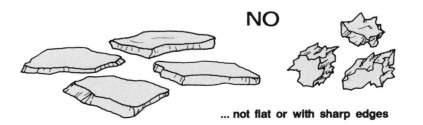

... not flat or with sharp edges

11. The aggregates should be **clean** and free from dirt and organic material. As with sand, you should wash them if necessary (see Section 33, paragraph 8).

12. Gravel and broken stones usually range in size from 0.5 to 6 cm across. For particular construction works, such as relatively thin concrete walls and slabs, you will have to use smaller broken stones.

Note: the sizes of the largest aggregate particles should never exceed one-quarter of the concrete thickness.

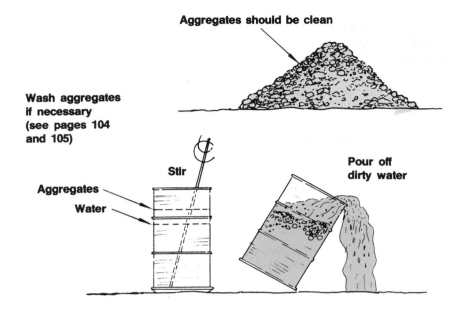

Aggregates should be clean

Wash aggregates if necessary (see pages 104 and 105)

Stir

Aggregates

Water

Pour off dirty water

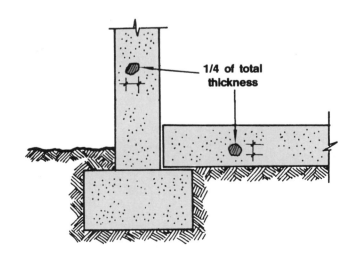

Aggregates should never exceed 1/4 of the concrete thickness

1/4 of total thickness

13. For heavier concrete work, particularly base slabs and heavy foundations, large boulders and rocks may be thrown in, provided the concrete can be packed around and over them.

14. In areas where rock aggregate is unavailable, broken brick is frequently used. It does not make strong concrete but can be acceptable for simple foundations and lightly loaded walls. Care should be taken in preparing, placing and curing the concrete to make it as strong as possible.

**Large rocks may be used for base slabs
If they do not exceed
1/4 of the slab thickness**

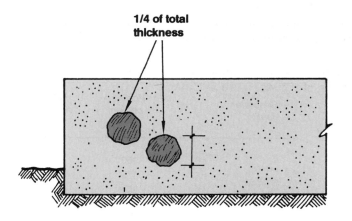

1/4 of total
thickness

**You can use broken bricks
for simple foundations**

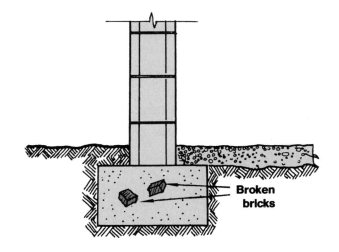

Broken
bricks

TABLE 9

Amount of materials required to prepare 1 m³ of concrete

Concrete quality*		Portland cement (kg)	Sand 0.2-5 mm (l)	Gravel 5-25 mm (l)	Broken stones 25-60 mm (l)	Approx. water (l)	Examples of constructions
Lean C7-C10*	(a)	150**	400	—	800	100	Subfoundation
	(b)	175**	375	1 000	—	—	Subfoundation, forms
	(c)	200**	400	600	300	150	Forms
Ordinary C15-C20	(a)	250**	300	1 000	—	—	Foundation water control structure,
	(b)	250**	400	600	300	170	floor slabs
Rich C25-C35		350**	450	800	—	200	Underwater foundation, reinforced concrete (monk, fishing pit, spillway, etc.)
Very rich C40-C60		400***	500	750	—	—	Water pipes, canals

* "C" refers to approximate concrete strength in newtons/mm²
** Quality 250
*** Quality 325, if available

15. For general-purpose concrete work, there are three basic ways to determine the correct proportions of aggregates and cement:

- simple rules of thumb, based on the kind of construction and defining the **weight of cement per cubic metre of concrete**;
- simple rules of thumb, based on the kind of construction and defining the **ratios by volume**;
- the more accurate method, based on the **free-pore volume** inside the concrete.

16. For small concreting jobs and for repairs, use one of the first two methods. For larger concreting jobs, it is safest to use the third method.

17. The simple rules of thumb are given as guidelines for the preparation of four basic types of concrete, from a lean to a very rich mix.

18. **Table 9** provides guidelines based on the weight of the cement for concretes containing from 150 to 400 kg cement per m³. The quantity of water to be used depends greatly on the moisture content of the sand and gravel and must be judged when mixing the concrete.

19. **Table 10** provides guidelines for similar types of concrete based on ratios by volume. The water quantity to be used is about 0.75 l per litre of cement. This should be checked while mixing.

TABLE 10

Proportions of concrete materials by volume

Concrete quality	Proportion by volume				
	cement	:	sand	:	gravel
Lean	1	:	4	:	6
	1	:	3	:	5
Ordinary	1	:	2	:	4
	1	:	2	:	3
Rich	1	:	2	:	2
Very rich	1	:	1.5	:	2.5

Note: these figures (e.g. 1:2:4) are commonly used to describe concrete mixes; the weight mixes in **Table 9**, however, give more accurate specifications

20. The free-pore volume method is based on the fact that the cement should fill the pore spaces left free in the aggregates.

21. The volume of these free-pore spaces and the amount of cement paste required can be determined as follows.

(a) Take a sample of the ungraded aggregates you will be using to prepare your concrete.

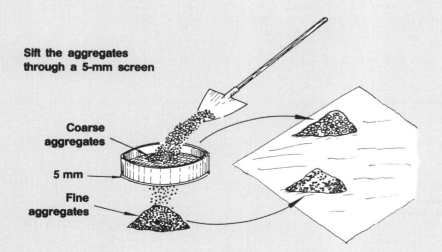

Take a sample of aggregates

(b) Sift this sample through a 5-mm screen to divide it into coarse aggregates (larger than 5-mm diameter) and fine aggregates (smaller than 5-mm diameter).

Sift the aggregates through a 5-mm screen

Coarse aggregates

5 mm

Fine aggregates

(c) Fill a small container of volume **V1**, such as a 15-litre bucket, with dry, coarse aggregates.

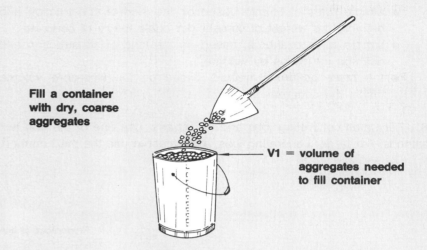

Fill a container with dry, coarse aggregates

V1 = volume of aggregates needed to fill container

(d) Add water to the container and measure the water volume **V2** required to fill it, in litres.

(e) This volume equals the volume of fine aggregates and cement needed to fill the free-pore space of the coarse aggregates.

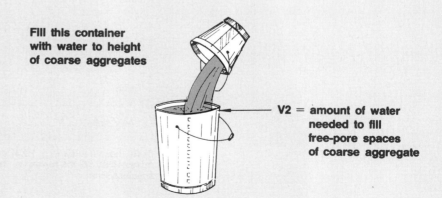

Fill this container with water to height of coarse aggregates

V2 = amount of water needed to fill free-pore spaces of coarse aggregate

(f) Measure a volume **V2** of fine aggregates equal to the water volume determined in the previous step, and put these fine aggregates into another container.

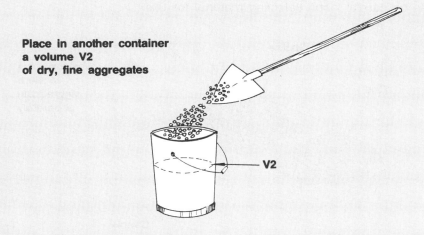

Place in another container a volume V2 of dry, fine aggregates

— V2

(g) Slowly add water to this container and measure (in litres) the water volume **V3** required to bring the water level to the top of the fine aggregates. This volume equals the volume of cement needed to fill the remaining free-pore space after mixing all aggregates.

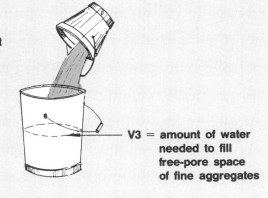

Fill this container with water to height of fine aggregates

V3 = amount of water needed to fill free-pore space of fine aggregates

(h) Add 10 percent to this volume to obtain the corrected volume **V4** of cement paste.
(i) Divide **V1** by **V4** to obtain **A**.
(j) Divide **V2** by **V4** to obtain **B**.
(k) Add **A** to **B** to obtain **C**.
(l) The ratio of cement to ungraded aggregates, by volume, should be **1:C**. One part of cement should be used for **C** parts of this particular quality of aggregates.

22. To determine the approximate amount of water required according to the type of concrete:

- very rich concrete: about 24 l of water per 50-kg cement sack;
- rich to ordinary concrete: about 28 l of water;
- lean concrete: about 33 l of water.

Example

You are using a 20-litre bucket.

(a) Fill this bucket with coarse aggregates: **V1** = **20 l**.
(b) It takes 13.3 l of water to fill the pore spaces of these coarse aggregates: **V2** = **13.3 l**.
(c) Put 13.3 l of fine aggregates in a second container.
(d) It takes 6.2 l of water to bring the water level to the top of these fine aggregates: **V3** = **6.2 l**.
(e) Add 10 percent to **V3** to obtain **V4** = 6.2 + 0.62 = 6.82 = **6.8 l**.
(f) Determine **A** = **V1** ÷ **V4** = 20 ÷ 6.8 = **2.94 l**.
(g) Determine **B** = **V2** ÷ **V4** = 13.3 ÷ 6.8 = **1.96 l**.
(h) Determine **C** = **A** + **B** = 2.94 + 1.96 = **4.9 or 5 l**.

Therefore, in this example the ratio of ingredients should be one part of cement to five parts of ungraded aggregates.

Note: you will have to recalculate **C** if you use a different type of aggregates.

Measuring the concrete components

23. To measure accurately the amounts of cement, sand and gravel or stones required to prepare good concrete, you can use one of the methods described earlier (see Section 33).

24. If your sand is very dry, moisten it a little before measuring the amount required.

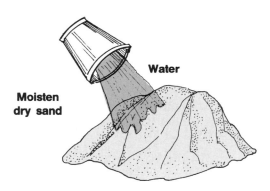

Moisten dry sand

Water

25. For the measurement of volumes of ungraded aggregates to be based on a cement volume, it is useful to use a homemade box containing 40 l, the approximate volume of one 50-kg cement bag.

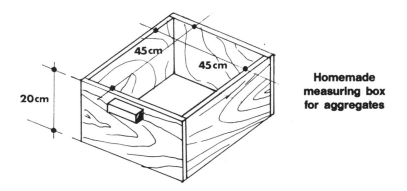

45 cm

45 cm

20 cm

Homemade measuring box for aggregates

Storing concrete components

26. You can store aggregates in piles or bins, but be careful not to let the separate sizes mix in with each other. Store them in separate areas, or use a wooden divider between each grade. Remember also that after some time, the bigger sizes tend to be at the bottom and sides of the pile, so be careful when selecting material for use.

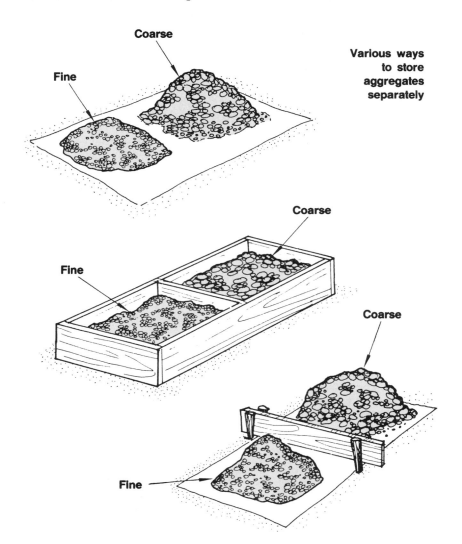

Various ways to store aggregates separately

Coarse

Fine

Coarse

Fine

Coarse

Fine

27. To prepare concrete by hand, you need a clean, watertight mixing area. Small amounts of concrete may be mixed on level ground using:

- **a metal sheet** (1 × 2 m);
- **a wooden platform** (1 × 2 m) carefully built to be watertight;
- **a portable mixing platform** made from galvanized sheet metal (about 1 × 2 m) nailed on two planks (5 × 30 × 180 cm). The curved ends are nailed vertically to be watertight.

28. Larger amounts of about 50 kg of cement may be mixed using:

- **a wooden platform** (2 × 3 m) carefully built to be watertight;
- **a concrete mixing floor:** if the floor is 5-cm thick with a diameter of 250 cm and a 10-cm raised edge, you will need about 200 l of lean concrete to build it. For example, mix 50 kg of cement, 120 l of sand, 300 l of gravel (5- to 12-mm diameter) and about 30 l of water.

**For mixing
small amounts of concrete**

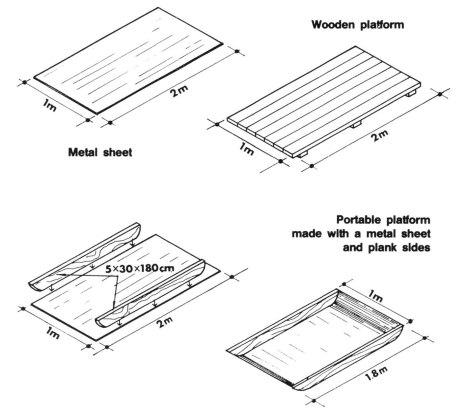

Wooden platform

Metal sheet

Portable platform
made with a metal sheet
and plank sides

5×30×180 cm

**For mixing
large amounts of concrete**

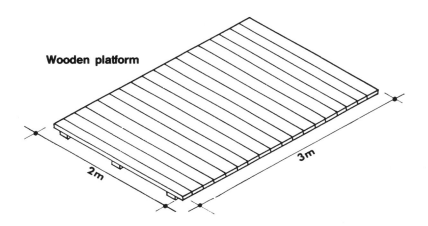

Wooden platform

2m

3m

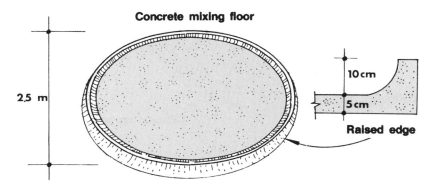

Concrete mixing floor

2,5 m

10 cm

5 cm

Raised edge

121

29. Determine how much of each ingredient you will require to prepare a certain amount of concrete and then proceed as follows.

(a) **Pour the sand** on to the mixing area and **spread it** evenly.

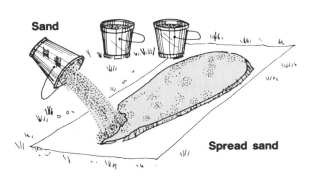

Sand

Spread sand

(b) **Spread the cement** evenly over the sand.

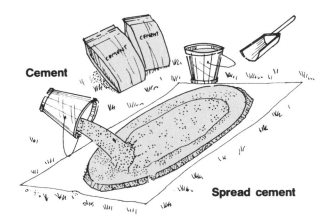

Cement

Spread cement

(c) **Mix** the cement and sand well, stirring with a shovel, until you obtain a uniform colour; **spread** this mixture evenly over the mixing area.

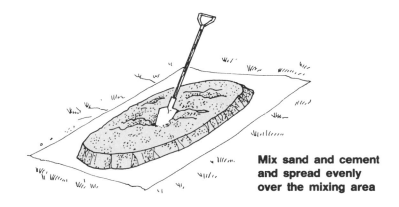

Mix sand and cement
and spread evenly
over the mixing area

(d) **Wet the gravel** and **spread** it evenly over the mixture.

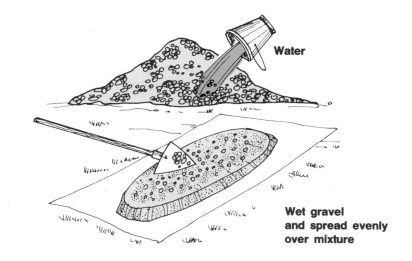

Water

Wet gravel
and spread evenly
over mixture

(e) **Mix thoroughly** together to obtain a homogeneous mixture.

Mix gravel

(f) **Rake into a pile and form a hollow** in the middle of the mixture.

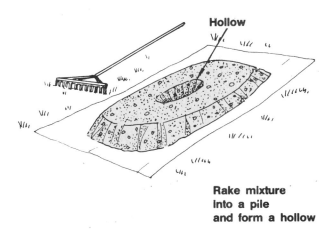

Hollow

**Rake mixture
into a pile
and form a hollow**

(g) From the previously measured volume, **slowly add water** in the centre and progressively moisten the mixture.

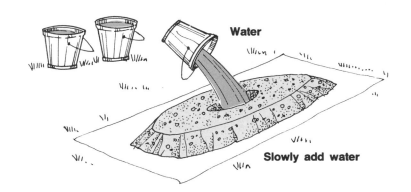

Water

Slowly add water

(h) **Shovel back and forth**, mixing thoroughly until you obtain concrete of uniform plastic consistency.

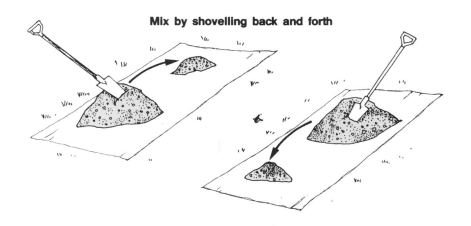

Mix by shovelling back and forth

(i) If the concrete is either too wet or too dry, correct its consistency (see paragraph 32 of this section).

Preparing good concrete mechanically

30. If a concrete mixer is available, it will be much easier to prepare concrete. The quality of the concrete is also likely to be better. As the capacity of concrete mixers may vary from 150 l to 500 l or more, it is important to select a machine adapted to your needs. You should know the capacity of your mixer and plan the concrete mixing accordingly.

31. Before preparing a concrete batch, gather all the ingredients required close to the mixer. Then proceed as follows.

(a) Pour 10 percent of the water required in the drum.
(b) Add half of the coarse aggregates, gravel and/or stones.
(c) Start mixing.
(d) Add all the cement required for the batch.
(e) Wait for 30 seconds.
(f) Add all the sand required.
(g) Add the rest of the water.
(h) Add the rest of the coarse aggregates.
(i) Mix for 4 minutes.
(j) Check the consistency and correct it if necessary (see paragraph 32 in this section).

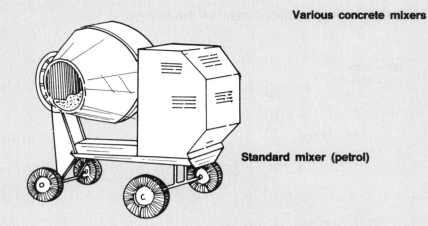

Various concrete mixers

Standard mixer (petrol)

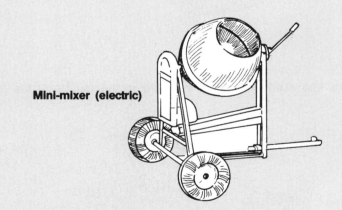

Mini-mixer (electric)

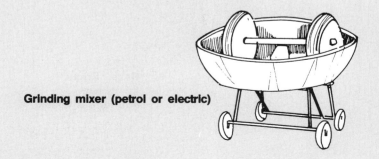

Grinding mixer (petrol or electric)

Note: when mixing concrete by hand or machine:

- beware of dry cement losses on windy days;

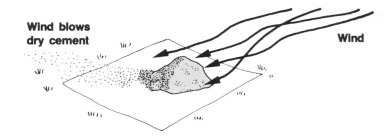

Wind blows dry cement

Wind

- do not use too much water;
- when mixing by hand, do not wash cement out with the water;

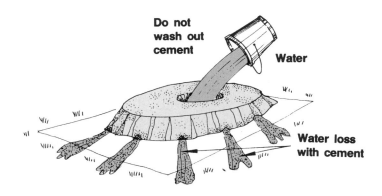

Do not wash out cement

Water

Water loss with cement

- rinse and clean your tools and your mixer well at the end of each working session.

Clean and rinse tools well

Correcting the consistency of concrete

32. Good fresh concrete should have a plastic consistency. If this is not the case, its consistency should be corrected as follows:

- if the mix is **too wet**, add small amounts of sand and gravel in the right proportion until the mix becomes plastic;

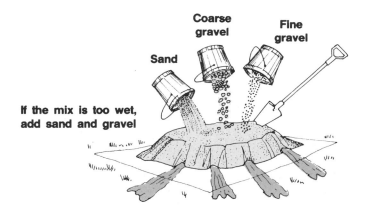

Coarse gravel

Fine gravel

Sand

If the mix is too wet, add sand and gravel

- if the mix is **too dry**, add small amounts of water and cement in the right proportions until the mix becomes plastic.

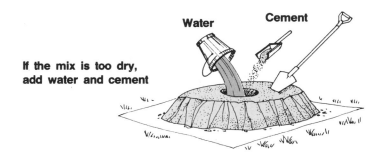

Water

Cement

If the mix is too dry, add water and cement

33. Take note of the amounts of the materials added so that you will have the corrected proportions for the next batch of concrete.

125

34.   For larger construction works or where the strength is critical, the quality of fresh concrete should be routinely tested before use. This can be simply done using the slump test, which provides a relative measurement of the fresh concrete plasticity and its expected strength after setting or hardening.

35.   To execute the slump test, you will need:

- **a conically shaped bucket** (15 to 20 l);
- **a wooden rod** about 60 cm long and 15 to 20 mm in diameter with well-rounded ends;
- **a base plate** at least 30 × 30 cm, either a thick wooden board or preferably a steel plate.

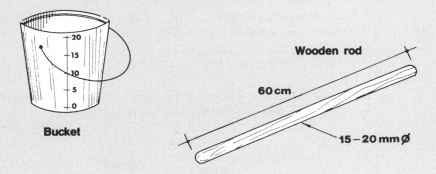

Bucket

Wooden rod

60 cm

15 – 20 mm ∅

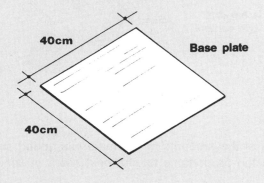

40 cm

Base plate

40 cm

36.   Proceed as follows, using freshly mixed concrete.

(a)   **Wet the bucket** and the base plate.

Wet bucket and plate

Water

(b)   **Fill the bucket** with the concrete to be tested, by placing it in layers of about 10 cm.

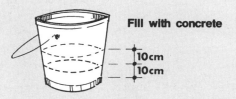

Fill with concrete

10 cm
10 cm

(c)   Use the wooden rod to **pack each layer** thoroughly before filling the next layer.

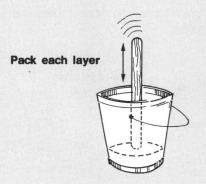

Pack each layer

(d) **Smooth the surface** of the concrete to enable you to fill the bucket exactly to the top.

Smooth surface

(e) Carefully **turn the bucket upside down** on to the base plate.

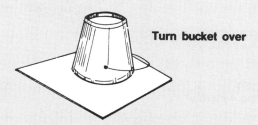

Turn bucket over

(f) Carefully **lift the bucket off** the concrete, place it alongside and immediately measure, in cm, the difference between the height of the slump cone and the height of the bucket (original concrete cone).

(g) This difference is called **the slump**.

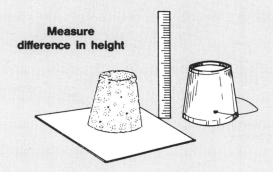

Measure
difference in height

37. Compare the measured slump with the range of values suggested according to the type of construction (see **Table 11**). Usually, a slump of 25 to 30 percent should be taken as satisfactory. Note that with some standard mixes, the concrete may shear. If so, repeat the test, or estimate the slump from the upper edge of the remaining sample.

TABLE 11

Slump test: acceptable ranges of slump*

Type of construction	Slump range
Slabs and thin reinforced structures	25–50%
Work floors, culverts and drainage structures	10–25%
Walls without reinforcement	10–35%
Walls with reinforcement	20–50%

* Expressed as percentage of bucket height at 30°C

38. If the slump is not satisfactory, the cement quality has to be improved using the same proportions as in the original mix as follows:

● **to decrease the slump**, add sand and gravel;
● **to increase the slump**, add water and cement.

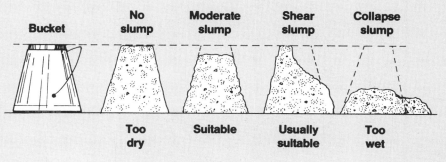

127

39. Concrete is usually used in conjunction with forms (**shutterings**), which determine the final shape of the concrete structure to be built. In many cases the concrete is reinforced (see Section 35).

40. The forms in which the concrete can be cast are generally made of light, cheap wooden boards and timber pieces, nailed or bolted together. For a series of standard shapes, steel plate is sometimes used.

41. Good forms should have the following characteristics. They should:

- be **rigid** enough not to lose their shape when full of concrete;
- be **watertight**;
- be **easy to remove** without damaging the concrete;
- be **re-usable**, in case other similar structures have to be built;
- easily accommodate reinforcement material, if used.

42. Forms should be well braced so that they remain firmly in place.

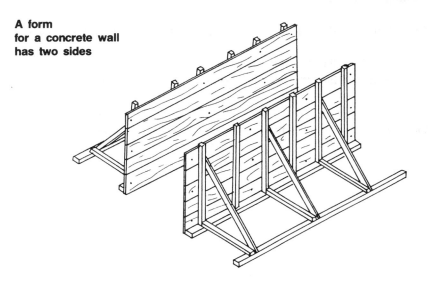

**A form
for a concrete wall
has two sides**

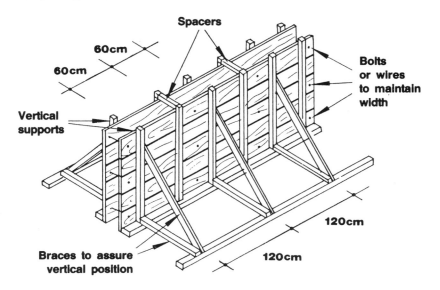

**Form ready
for placing concrete**

Spacers

60cm

60cm

Bolts
or wires
to maintain
width

Vertical
supports

Braces to assure
vertical position

120cm

120cm

Placing concrete

43. Concrete should be as fresh as possible when placed, ideally:

- **within 25 minutes** after opening the cement sack;
- **within 20 minutes** after adding water to the mix.

44. Once the concrete has started to set, it cannot be used. It is therefore important to have everything ready for use. Only make as much concrete in each batch as you can place in the time available.

45. Avoid placing concrete under water, because it is very difficult to make good concrete in these conditions. Use a drainage ditch, if necessary, to make sure that the concreting site is well drained. The earth, however, should be slightly moist. The supporting base should be firm and in many cases may require a layer of rock, brick rubble or other aggregate. If the concrete is to be firmly placed on rock, it should be well cleaned and dried.

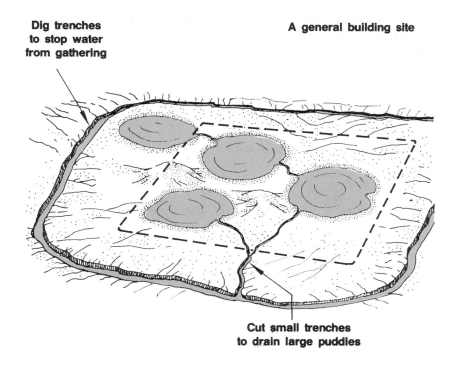

Dig trenches to stop water from gathering

A general building site

Cut small trenches to drain large puddles

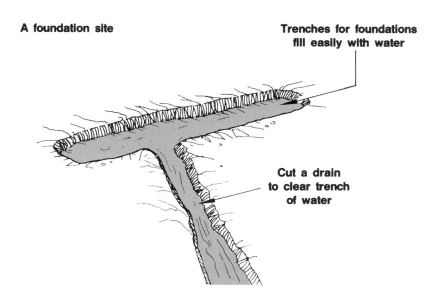

A foundation site

Trenches for foundations fill easily with water

Cut a drain to clear trench of water

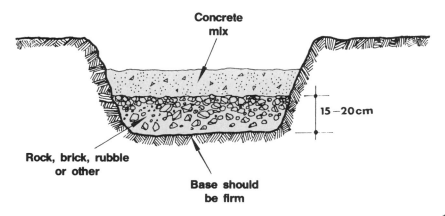

Placing concrete

Concrete mix

15 – 20 cm

Rock, brick, rubble or other

Base should be firm

129

46. Avoid segregating the ingredients of the concrete during placing, as this weakens the concrete and makes poor surfaces and poor joints between layers:

● never let concrete drop freely more than 1.5 m;
● never let it slide down a very steep incline;
● do not transport it over a long distance without remixing it.

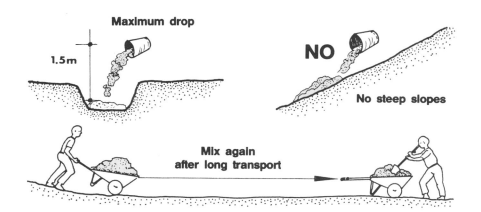

Maximum drop

1.5m

NO

No steep slopes

Mix again after long transport

47. Before placing the concrete in the forms, you should **oil or grease** their inside surface to make it easy to remove them once the concrete has set. You should also **wet the forms**.

48. Place concrete in layers 15 to 20 cm thick. The concrete should be strongly rammed to tighten the coarse aggregates and should have a 2- to 3-cm "soft" upper surface to ensure a strong liaison with the next layer.

49. Use a shovel, a wooden pole or a 2-cm diameter iron rod to tamp the fresh concrete firmly into place.

50. **Do not try to use "wet" concrete** to make better joints, as it will only separate more and the water will wash out, leaving a very poor surface and joint.

130

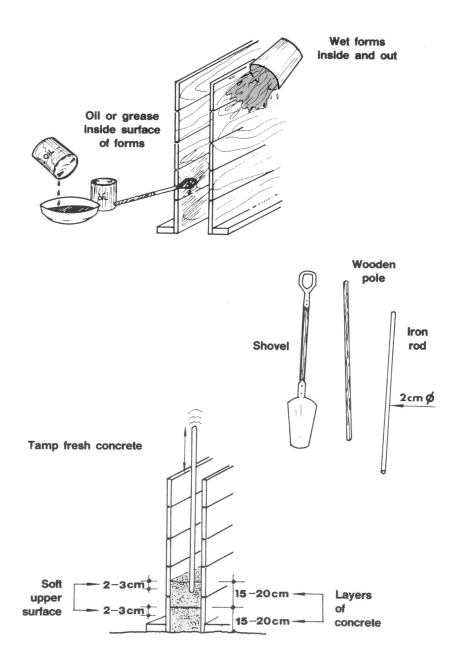

Wet forms inside and out

Oil or grease inside surface of forms

Shovel

Wooden pole

Iron rod

2cm ∅

Tamp fresh concrete

Soft upper surface — 2–3cm / 2–3cm

15–20cm / 15–20cm

Layers of concrete

51. You can also hammer on the outside of the forms to help to settle the concrete along the sides.

52. If the previous layer has set, **roughen its upper surface** to make a good connection to the next layer. You can also brush on a coat of liquid cement, which is cement dissolved in water. If available, use a cement bonding compound instead.

Settle concrete in forms by hammering

Note: it is always best to build a structure continuously, without interrupting the placing of the concrete. Remember the higher the structure, the stronger the forms have to be. If this is a problem, it may be necessary to build the structure in stages, letting each stage set before placing the new stage.

Roughen the surface

Coat with liquid cement or bonding compound

53. Within **less than half an hour** after adding water to the cement, the chemical reaction between these two ingredients results in the setting and progressive hardening of the concrete. The concrete acquires its strength, durability and impermeability during the curing process. To get the strongest possible concrete, curing should not be too rapid. It normally takes at least 28 days.

54. If the concrete is allowed to dry out, it will stop hardening; the curing process will not start again when the concrete is rewetted. Therefore, as soon as the concrete is placed, you should protect it from drying too fast by following these guidelines.

(a) Do not let the concrete dry out before placing it.
(b) Avoid placing concrete during the hot hours of the day.
(c) Wet the forms abundantly before placing the concrete. Keep them wet and do not remove them too quickly.
(d) Protect concrete from sun and wind, covering it with wet burlap, canvas, empty cement sacks, palm leaves, banana leaves, damp sand.
(e) Keep these covers wet so that they do not absorb water from the concrete.
(f) Sprinkle the concrete regularly with water after it has hardened enough not to wash out.

55. Keep the concrete from drying out:

- for at least seven days in temperate climates;
- for at least 11 days in warm climates.

56. It is best not to remove the forms from the concrete until the curing process has progressed well for at least 48 hours. In some cases, it may take as long as 21 days before the forms can be totally removed. When the forms have been removed, clean up any rough surfaces and fill any larger gaps or holes with mortar, if necessary.

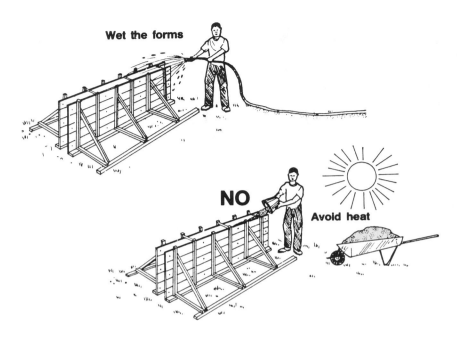

Wet the forms

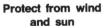

NO   Avoid heat

Protect from wind and sun

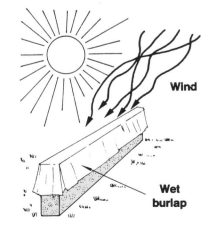

Wind

Wet burlap

Sprinkle until hard

Making concrete blocks

57. You can make simple concrete blocks using a standard wooden form, which can be reused (see page 100 for standard block sizes). A mix of 1:2:4 to 1:5:8 can be used, with aggregates smaller than 13 mm across and a fairly moist mix. You will need to cure the blocks carefully as they will split and break if handled too soon or left to dry out. Normally, the finish is not as good as that for machine-made blocks, which are formed under pressure using a dry concrete mix.

Details for building the outer form and the inner mould for making a hollow block

Outer form

A simple box form for making a solid block

Note: the forms shown on this page are for blocks of 20 × 20 × 40 cm

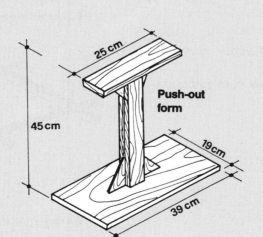

Push-out form

Inner form

A two-piece form for making a hollow block

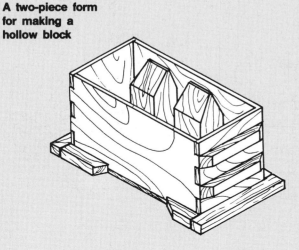

When the cement is nearly dry, carefully remove the inner form and slowly push the block from the outer form

133

35   Concrete reinforcement

1.   Reinforced concrete is made by adding reinforcement to normal concrete. Reinforcement is used to keep the concrete from collapsing.

Selecting reinforcement

2.   There are three main kinds of concrete reinforcement:

- round steel bars, with a standard diameter from 5 mm to 40 mm;
- diamond mesh, "expanded metal" used for reinforcement of light concrete **slabs*** (the longer mesh diagonal is set perpendicular to the slab supports);
- welded wire mesh with rectangular mesh of standard dimensions.

3.   To use reinforcement, you may also need:

- tie wire, soft annealed steel wire from 0.7 to 1 mm in diameter, to connect bars and mesh together;
- spacers, in metal, wood, plastic, etc., to make sure reinforcement is positioned correctly within the forms.

Round steel bars

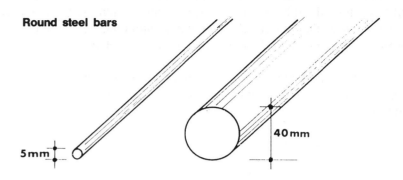

5mm   40mm

Expanded metal mesh

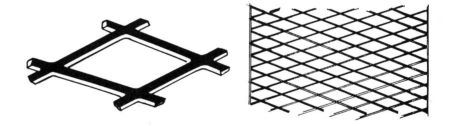

Welded wire mesh

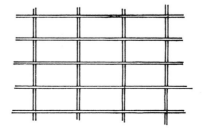
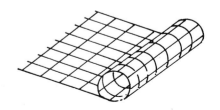

Using reinforcement

4.   The amount of reinforcement needed for a particular construction should be determined by an engineer, who should also specify how and where the reinforcement should be placed within the concrete to avoid collapse. Simple specific designs are given in **Pond construction, 20/2**.

5.   For your guidance only, the number of steel bars is generally calculated as a percentage of the gross area of each section of concrete as follows:

- for foundations: at least 1 percent;
- for slabs: at least 3 percent;
- for columns: at least 6 percent.

Area of steel (in mm²) present in concrete section according to diameter and number of steel bars* (general formula = 3.1416 × d² × n ÷ 4)

Diam. (mm)	Weight (kg/m)	Cir- cumf. (mm)	Number of steel bars										
			1	2	3	4	5	6	7	8	9	10
4	0.098	12.57	12	25	37	50	62	75	88	100	118	125
5	0.154	15.70	20	39	59	78	98	118	138	157	176	196
6	0.222	18.84	28	56	85	113	141	170	198	226	294	282
7	0.302	21.98	38	76	115	153	192	230	269	307	346	384
8	0.395	25.14	50	100	151	201	251	301	352	402	452	502
9	0.499	28.28	63	127	190	254	318	381	445	508	572	636
10	0.617	31.42	79	157	236	314	393	471	550	628	706	785
11	0.746	34.55	95	190	285	380	475	570	635	760	855	950
12	0.888	37.71	113	226	339	452	565	679	792	904	1 017	1 131
13	1.042	40.80	132	265	398	530	663	796	929	1 061	1 194	1 327
14	1.208	43.99	154	308	462	616	770	924	1 078	1 231	1 385	1 539
15	1.387	47.10	177	353	530	707	884	1 060	1 237	1 413	1 590	1 767
16	1.578	50.10	201	402	603	804	1 005	1 206	1 407	1 608	1 809	2 010
17	1.782	53.40	226	453	680	907	1 134	1 361	1 588	1 815	2 042	2 269
18	1.998	56.54	254	509	763	1 018	1 272	1 526	1 780	2 035	2 290	2 544
19	2.226	59.70	283	567	850	1 134	1 417	1 701	1 984	2 268	2 551	2 835
20	2.466	62.82	314	628	942	1 257	1 571	1 884	2 199	2 513	2 827	3 141
25	3.853	78.60	491	982	1 473	1 963	2 454	2 945	3 436	3 926	4 417	4 908
30	5.549	94.30	707	1 414	2 121	2 827	3 534	4 241	4 948	5 654	6 361	7 068
32	6.313	100.50	804	1 608	2 413	3 217	4 021	4 826	5 630	6 434	7 238	8 042
35	7.553	110.01	962	1 924	2 886	3 848	4 811	5 773	6 735	7 696	8 659	9 621
40	9.865	125.70	1 256	2 513	3 770	5 026	6 283	7 540	8 797	10 053	11 309	12 566

* For round and smooth steel bars

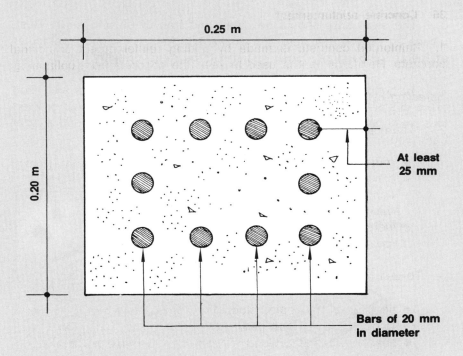

0.25 m

0.20 m

At least 25 mm

Bars of 20 mm in diameter

Example:

A reinforced concrete column 0.20 × 0.25 m is to be built. The necessary reinforcement can be estimated as follows:

(a)   Calculate the gross area of the section of the column: 0.20 m × 0.25 m = 0.05 m² = 500 cm²

(b)   Calculate the minimum area of the steel reinforcement required: 500 cm² × 0.06 = 30 cm² = 3 000 mm²

If you plan to use 10 steel bars as shown in the drawing above, start with the 10-bar column on the right-hand side of the chart on this page. Follow the column down until you find an area at least equal to 3 000 m² or in this case 3 141 mm². Now follow this line across and you will see that this area corresponds to a steel bar diameter of 20 mm.
So, for this size of concrete column use reinforcement made of 10 steel bars of 20 mm.

Preparing the steel bar reinforcement

6.   The steel bars should be clean and free of oil and earth. Rust, unless so severe that bars are weakened, does not require particular attention, although any loose rust should be removed with a wire brush.

7.   To bend the steel bars according to design, you need a well-fixed piece of steel plate or heavy planking in which you have driven or fixed four small pegs of 10-mm diameter steel bar. If you plan to bend a lot of bars, you may prefer to build a strong workbench.

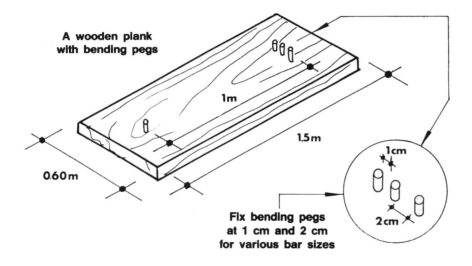

A wooden plank with bending pegs

1m

1.5m

0.60m

Fix bending pegs at 1 cm and 2 cm for various bar sizes

1cm

2cm

8.   Buy a special clamp or make one yourself by sawing a narrow notch in a piece of very thick steel bar or rod.

9.   Insert the steel bar to be bent between two of the first three pegs, making sure that the bar is positioned to bend in the right place. Using the clamp, bend the steel bar at the level of the single peg.

10.   Once the steel bars have been cut and bent according to design, the reinforcement is constructed. The bars should be firmly and securely bound together at their intersections with tie wire (see paragraph 3 on page 134).

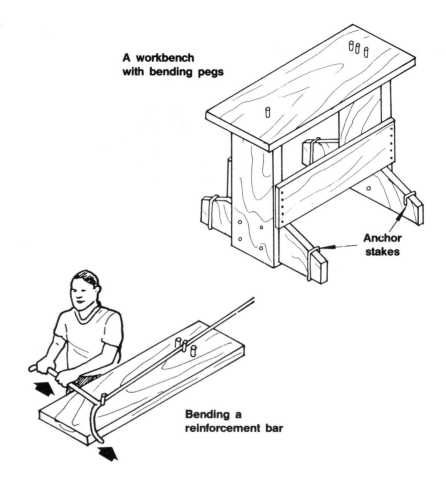

A workbench with bending pegs

Anchor stakes

Bending a reinforcement bar

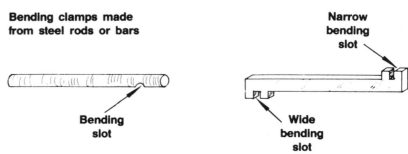

Bending clamps made from steel rods or bars

Narrow bending slot

Bending slot

Wide bending slot

Making reinforced concrete slabs

11. By using wire mesh reinforcement, you can make simple slabs with a fairly moist concrete mix of 1:2:4 to 1:5:8 that has aggregates smaller than 13 mm across. To make a slab, place the concrete inside a simple wooden form resting on a flat surface or level a piece of ground, cover it with a heavy plastic sheet and put the wooden form on that. As with blocks, you should take care in curing the concrete.

Note: the wire mesh may be held in place within the form using strips of wood over the top of the form and hang wires (see below). Remember to **leave at least 25 mm of free space all around the mesh and between the top and bottom of the form**. Also, it is often useful to set in one or more small loops to use as handles for lifting or moving the finished slab.

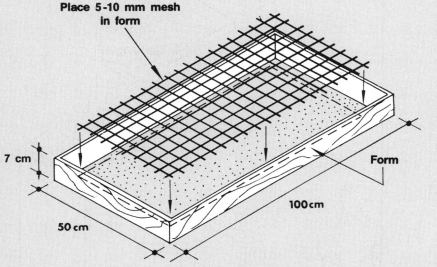

A reinforced concrete slab

Place 5-10 mm mesh in form

7 cm

Form

100 cm

50 cm

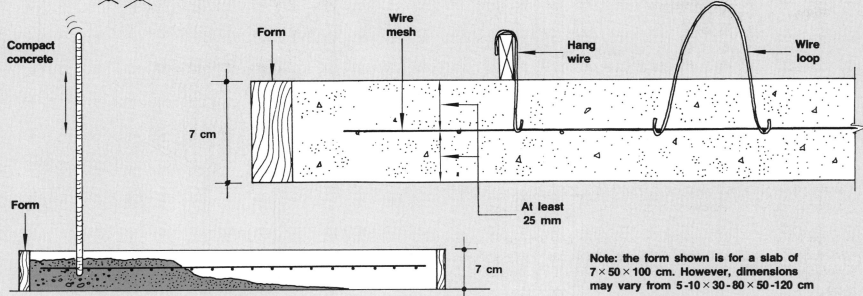

Detailed section through form with concrete in place

Compact concrete

Form

Form

7 cm

Wire mesh

Form

Hang wire

Wire loop

7 cm

At least 25 mm

7 cm

Note: the form shown is for a slab of 7 × 50 × 100 cm. However, dimensions may vary from 5-10 × 30-80 × 50-120 cm

Making reinforced concrete

12. To reinforce concrete, proceed as follows.

(a) **Fix the reinforcement well**, exactly according to the engineering design. There should normally be at least 25 mm between the bars and the outside surface. Make sure tie wires are sound, and that the bars are not twisted.

(b) **Place the forms** around the reinforcement. If necessary, use spacers to hold reinforcement bars in place.

(c) **Wet the form** and the reinforcement well.

(d) **Place the concrete** into the form, without disturbing the reinforcement.

A reinforced concrete column

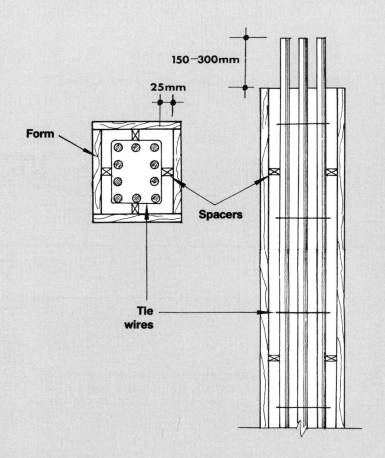

150—300mm

25mm

Form

Spacers

Tie wires

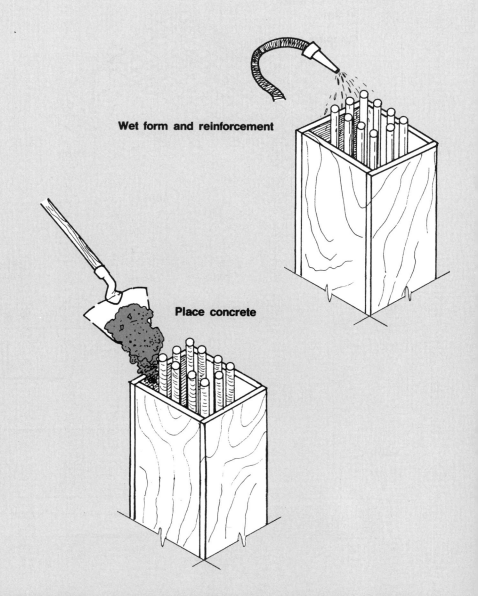

Wet form and reinforcement

Place concrete

(e) **Compact the concrete well**, especially around the reinforcement, but do not disturb the reinforcement or shake it.

(f) Take particular care in making **good joints** between layers.

(g) **Cure the concrete** well before removing the form.

(h) **Remove any spacers** if used, and finish and fill the outer surfaces. Be careful to make sure no reinforcement is exposed to water.

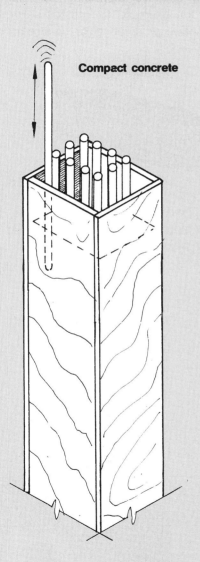

Compact concrete

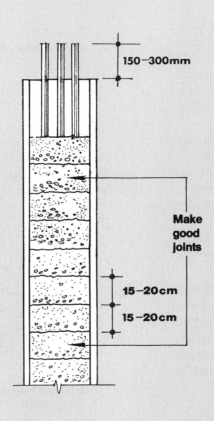

150–300mm

Make good joints

15–20cm

15–20cm

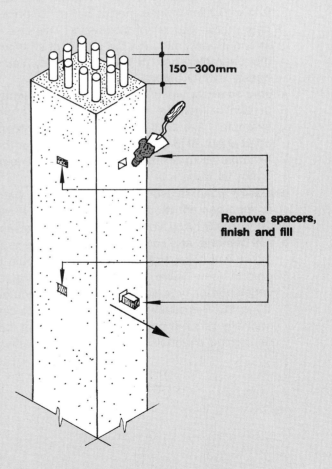

150–300mm

Remove spacers, finish and fill

36 Other construction materials

1. There are several other materials commonly used for construction purposes, particularly where supplies for normal cement or concrete are not easily available. These are generally not so strong or durable, but may be used if necessary. There are also many specialized materials, but these are usually too expensive or too complex to use for most fish culture constructions. Some of the materials you might use are:

- **lime mortars** which use **slaked lime***, produced from crushed and burnt limestone, mixed 1:2 to 1:3 with sand. You need about 0.15 to 0.2 m^3 of lime mortar per m^3 of brickwork and 0.2 m^3 of lime mortar per m^3 of stonework. This can make a reasonably strong mortar, although the quality of all lime mixes depends on the characteristics of the lime used. You should check the quality locally before using;
- **lime concrete** which can be made using **lime** instead of cement, in similar volume proportions as in ordinary concrete (for example 1:2:4, 1:3:6, etc.). As lime is lighter than cement, you need about 20 to 25 percent less weight for the same volume, for example 40 kg of lime instead of 50 kg of cement. Lime concrete is not as strong as cement concrete;
- **plaster** which is used for making smooth wall surfaces. Smoother than cement mortar, it is made with a range of mixes from 1:1:5 to 1:3:12 by volume of cement:lime:sand;
- **soil cements and concrete** in which cement (sometimes lime) is mixed with local soil. Ideally the soil should be reasonably well graded (see **Table 8** for suitable soil types) and free from vegetation or organic matter. Ratios of cement:soil are typically 1:3 to 1:6. The quality of soil cement varies considerably: it is not normally recommended for structural use, but may be used for reinforcing channels, tops of dikes or pathways.

37 Gabions

Introduction

1. A gabion is a **wire mesh cage or basket filled with stones**. Gabions are useful in construction works, for example to protect earth embankments, to line channels, to manage or divert river or stream flow and to protect river banks or coastlines.

Stone-filled gabion

2. You can buy wire mesh baskets and make your own gabions. The standard gabion basket consists of a single piece of wire mesh that can be assembled to form a rectangular box with a lid.

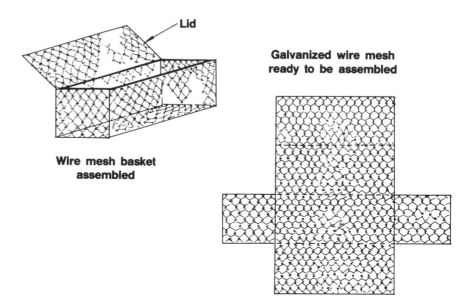

Lid

Galvanized wire mesh ready to be assembled

Wire mesh basket assembled

3. Wire mesh baskets for gabions can usually be found in two standard sizes. They are for:

- full-height one-metre gabions;
- half-height half-metre gabions.

4. The width of a standard basket is usually 1 m, and the length varies from 2 to 5 m or more.

5. Galvanized steel wire is used for making baskets for gabions. The wire is usually 3 mm in diameter and is twisted to form a mesh opening from 100 to 120 mm wide. Both single and double twist mesh are available, although double twist is better.

Standard gabion

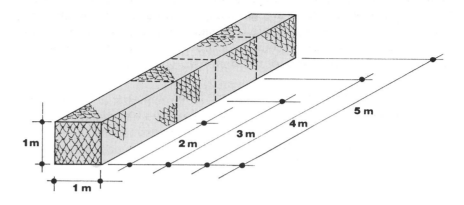

Half-height gabion

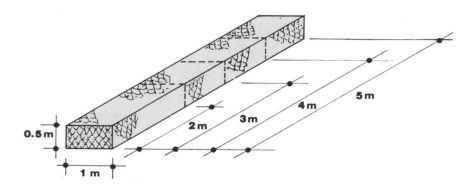

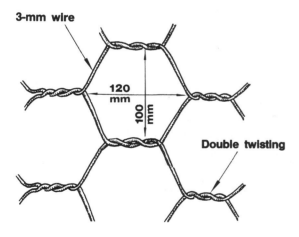

Advantages of gabions

6. Gabions have a number of important advantages in construction:

- **homogeneity and stability**, holding together and maintaining strong resistance to water current. Although containing small units (rocks, stones), each basket acts like a single large unit;
- **pliability**, adapting their shape easily to the ground contours even as these change gradually;
- **permeability**, allowing water to run through and act as filters for finer soil particles, thus giving protection to less stable materials;
- **simplicity** of design and easy, rapid construction;
- **economy**, using locally available stones that cost nothing.

Shape easily adapted

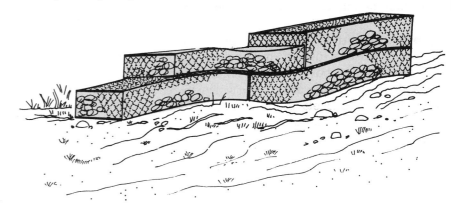

A three-gabion barrage

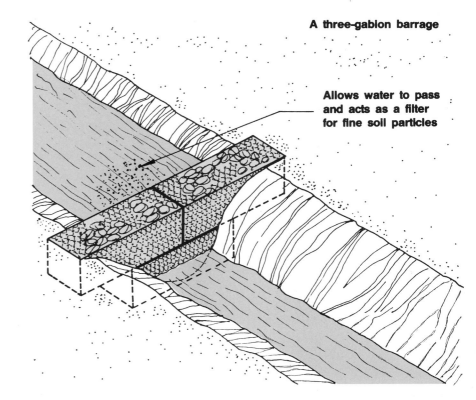

Allows water to pass and acts as a filter for fine soil particles

Designing gabion structures

7. Gabion structures normally consist of two parts:

    - **the foundation**, which must protect the structure against under-mining. It is usually made of half-height baskets and should extend well beyond the main structure body;
    - **the body**, which should resist the forces present. It is made of standard baskets of various sizes that are piled up in one or more rows according to the total height required.

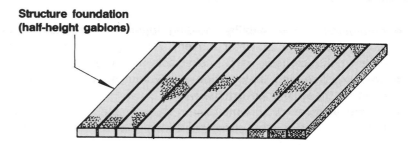

**Structure foundation
(half-height gabions)**

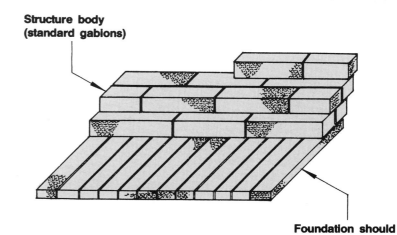

**Structure body
(standard gabions)**

**Foundation should
extend beyond
structure body**

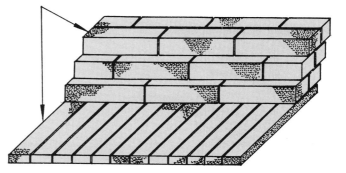

**Finished
gabion structure**

143

8.   Half-height gabions may also be placed on sloping river or stream banks or terraces. The gabions must be well supported at their base.

9.   For most fish farm uses, structures are not more than two or three gabions in height. Along stream banks or channels, a single gabion width is usually adequate. Structures two or three gabions or more in width may be needed for stream diversions in rapidly flowing water. Normally, the slope of gabion structures is from 45° to near vertical.

River or stream banks

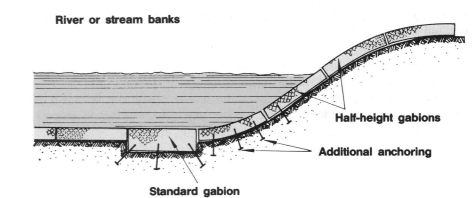

Half-height gabions

Additional anchoring

Standard gabion

River or stream diversion

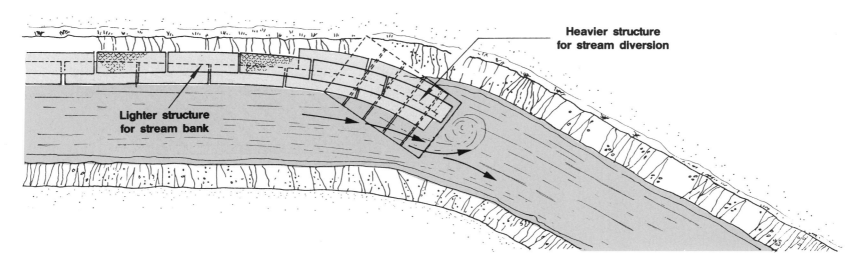

Heavier structure
for stream diversion

Lighter structure
for stream bank

River, stream or channel banks

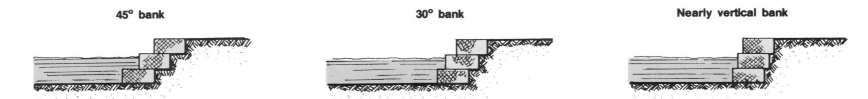

45° bank

30° bank

Nearly vertical bank

Building a gabion structure

10. Wire mesh baskets are built one at a time, put in place according to the design of the structure and then filled with stones. The following steps will tell you what to do.

(a) Begin the first basket by unfolding a section of wire mesh and stretching it out flat on the ground.
(b) Fold the front, back and sides to form a box with the lid open.
(c) Securely wire the four corners of the mesh box together as shown. This should be done very carefully using galvanized steel wire of the same quality and diameter as the mesh wire. Do not pull the wire with pliers, because you may tear and weaken the box.

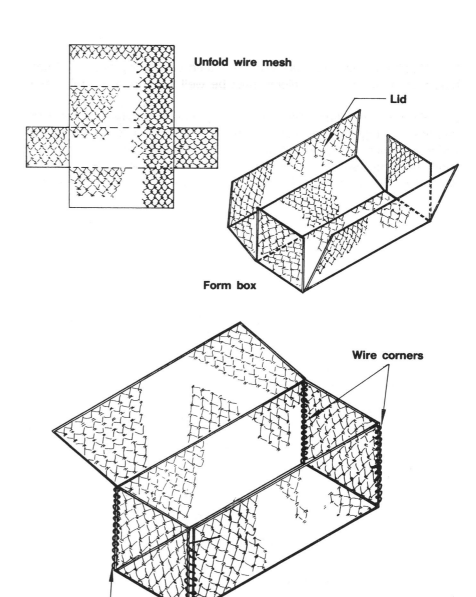

Unfold wire mesh

Lid

Form box

Wire corners

Wire corners

145

(d) After you have wired the four corners together, carry the basket to where you will use it.

(e) When a basket is in position, **make sure that it is straight and square**. To do this, stretch the front, back and side, driving a 1.5-m iron bar into each corner as shown. As each corner becomes straight and square, drive the bar into the ground to hold it in place.

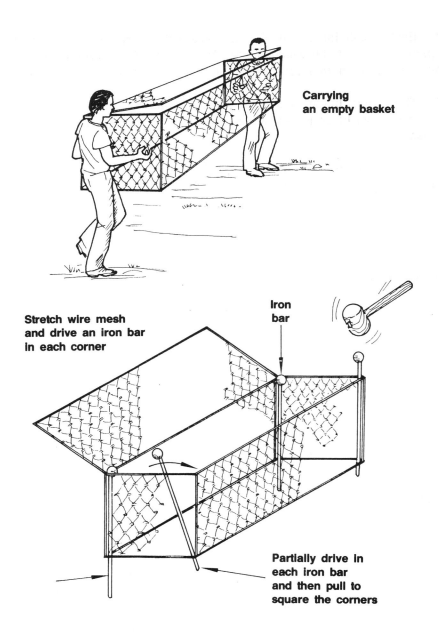

Carrying an empty basket

Stretch wire mesh and drive an iron bar in each corner

Iron bar

Partially drive in each iron bar and then pull to square the corners

11. Each mesh basket must also have **extra wire bracing** to help support the weight of the stones when the basket is filled. As braces, you can use the same wire you used to secure the four corners of the box.

12. The vertical braces are attached as soon as the basket is in place. The horizontal and angle braces are added as the baskets are filled with stones.

13. The drawings on this page show you where to put vertical, horizontal and angle braces, both for a full-height gabion and for a half-height gabion.

14. Each brace is attached by threading it through several of the wire mesh openings.

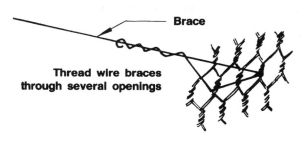

Brace

Thread wire braces through several openings

Braces for full-height gabion — 1 × 1 × 5 m

Top view

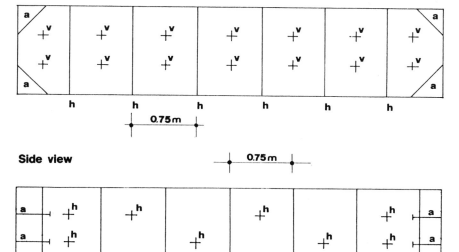

0.75 m

Side view

0.75 m

Braces for half-height gabion — 0.5 × 1 × 5 m

Top view

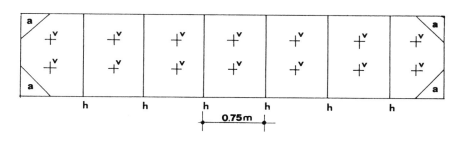

0.75 m

Side view

0.75 m

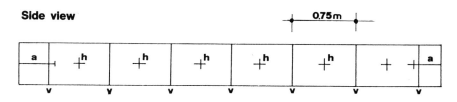

a = angle
h = horizontal
v = vertical

15. Now you are ready to start filling the basket with stones.

16. A **foundation basket** is best filled with round or rounded stones at least one-and-a-half times as large as the openings in the wire mesh. Avoid using stones larger than this. If the stones are too large, you cannot easily deform the basket to fit irregular or curved sites such as stream banks.

Note: try to choose stones that fit closely together so that there will be no large empty spaces in the basket.

17. A **structure basket** is also best filled with stones, at least one-and-a-half times as large as the openings in the wire mesh. However, if you do not have enough large stones, you can use smaller stones in the centre of the basket if they are at least 8 cm in diameter. If you use smaller stones, first line the bottom and sides with large stones, then fill the centre with smaller stones and finally cover the top with a layer of large stones.

Foundation basket

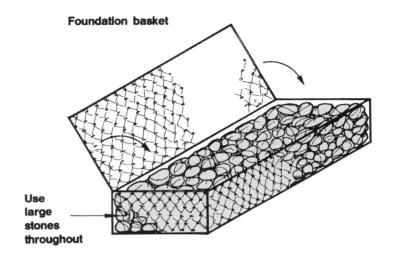

Use large stones throughout

Structure basket

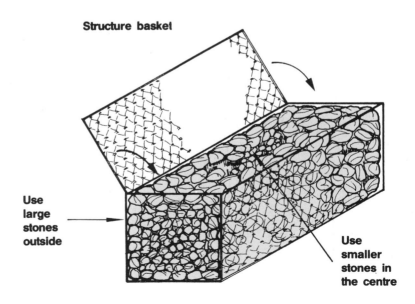

Use large stones outside

Use smaller stones in the centre

18. When you are filling baskets with stones, **make sure that the vertical wire braces stay vertical**.

19. Attach the **horizontal braces** and **angle braces** from time to time as you put in the stones.

Note: use hard stones such as granites, quartzites, sandstones, laterite and hard calcareous stones for filling baskets. Do not use schists, gneiss or serpentine, which are too friable, may break down in strong water currents and may eventually wash out of the baskets causing them to collapse.

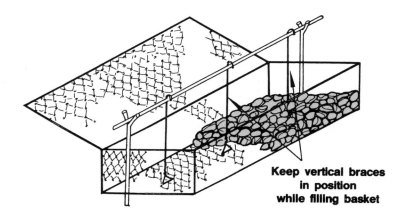

**Keep vertical braces
in position
while filling basket**

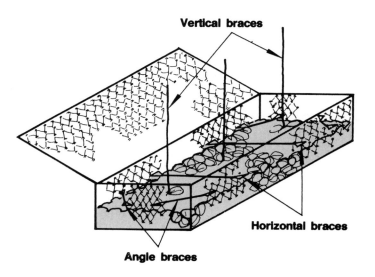

Vertical braces

Horizontal braces

Angle braces

20. When the basket is filled with stones, you can remove the iron bars from each corner.

21. Close the lid of the basket, pull the edges tight and fasten them every 20 cm with galvanized steel wire, using a short piece of iron for a lever as shown.

22. Finish the basket by attaching the vertical braces to the lid.

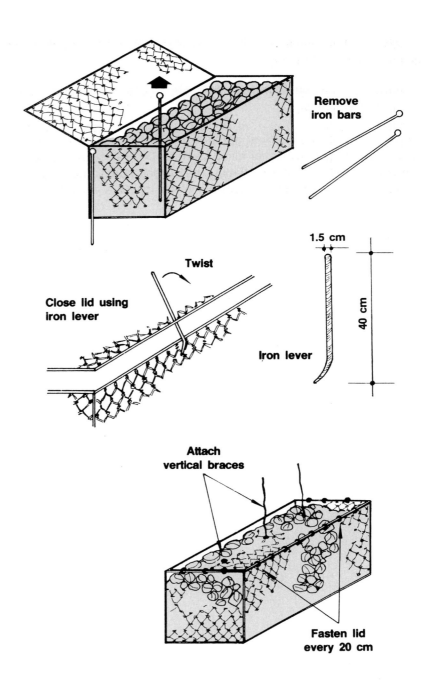

Remove iron bars

Twist

Close lid using iron lever

Iron lever

1.5 cm

40 cm

Attach vertical braces

Fasten lid every 20 cm

23. After the first basket is in place and filled, **add empty baskets** one by one according to the design of the gabion structure.

(a) Wire the back and sides of each new basket to the filled baskets already in position.
(b) Stretch the front corners of each empty basket using a 1.5-m iron bar until the basket is straight and square. Then hold it in place by driving the iron bar into the ground or into the gabion below.
(c) Attach the braces and fill the basket with stones as before. Remove the iron bars. Fasten the lid and attach the vertical braces.

24. Continue to add more empty baskets until the gabion structure is finished.

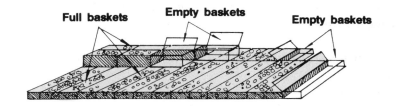

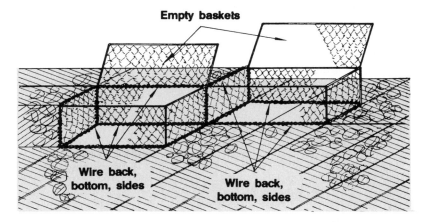

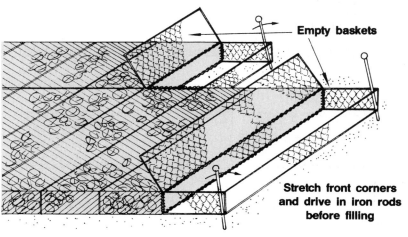

151

1.   Pipes are widely used on fish farms to transport water, for example through dams and dikes or under roads.

Most common types of pipes

2.   The type of pipe to be used depends on the size or diameter required for the planned water discharge capacity:

- **large pipes** (inside diameter 20 to 100 cm), concrete or ceramic;
- **small pipes** (inside diameter less than 20 cm), ceramic, plastic or galvanized. Bamboo pipes can also be used (see Section 31).

3.   Generally, concrete and ceramic pipes are cheaper for a given size, but cannot be made in small sizes.

Examples of concrete pipe

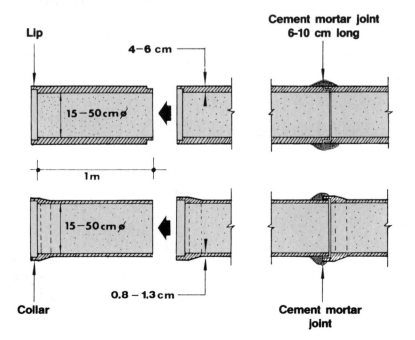

Selecting concrete pipes

4.   There are three types of concrete pipe. In order of increasing strength, they are:

- unreinforced concrete pipes;
- reinforced concrete pipes;
- asbestos-cement pipes.

5.   **Unreinforced concrete pipes** are usually precast, with a standard length of 1 m. They are connected by a joint sealed with cement mortar. Their diameter should be limited to a maximum of 50 cm. It is best to lay them under ground, at least 50 cm deep.

6.   **Reinforced concrete pipes** are rarely used in fish farming, perhaps only when very large diameters are needed.

7.   **Asbestos-cement pipes** are made by adding asbestos fibres to the concrete to increase strength. These pipes are more costly but have the advantage of being lighter, stronger and available in longer standard lengths (3 to 6 m). This reduces the number of joints to be sealed with cement mortar. The inside diameter usually varies from 15 to 30 cm. The pipes are laid down in a trench deep enough to protect them with at least 50 cm of soil. The foundation supporting them should be carefully built to accommodate the reinforced collars of the pipes.

Example of asbestos-cement pipe

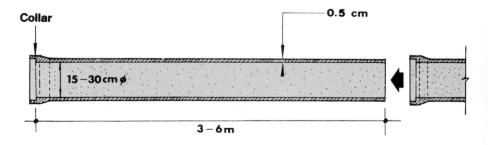

Selecting ceramic pipes

8. **Ceramic pipes** are made of baked clays, usually with a hard glazed exterior finish, with typical diameters of 10 to 20 cm. They are normally in short lengths, 50 cm to 80 cm, with a collar fitting at one end, sealed with mortar. Ceramic pipes are not strong and are easily broken in handling. As with plain concrete pipes, they must be well protected under ground.

Enlarged example of ceramic pipe

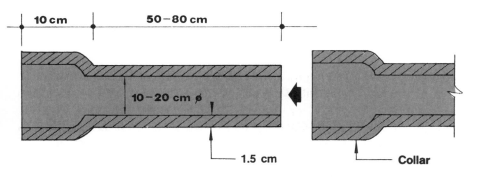

10 cm 50–80 cm

10–20 cm ø

1.5 cm Collar

Selecting galvanized pipes and plastic pipes

9. For smaller water flows, **galvanized iron pipes** (inside diameter 5 or 6 cm) or **plastic pipes** are normally preferred. The standard lengths available are usually longer (3 to 6 m), which reduces or may even eliminate the need for joints.

10. For plastic pipes, **pressure pipes** are stronger, heavier, and more expensive than **drainage pipes**. They are suitable for higher water pressures, for example pumped water supplies, and wall thickness depend on the pressure rating needed. **Drainage pipes** are lighter, have thinner walls, are cheaper and are suitable for low pressure, for example pond drains. The example below shows a drainage pipe with a flexible "o" ring inset in the collar for a "push fit joint".

11. It is best to protect plastic pipes from the sunlight, as they can become brittle if kept exposed.

Collar of plastic drainage pipe

Enlarged example of plastic drainage pipe showing detail of collar with flexible "o" ring

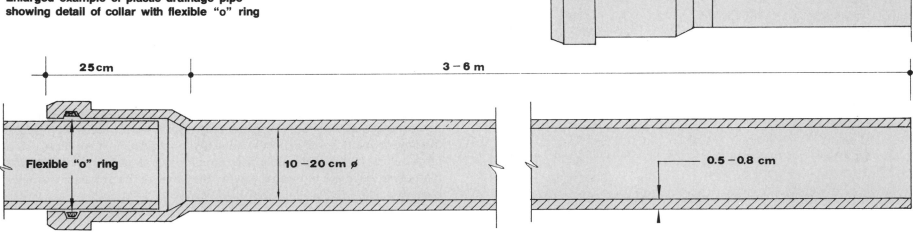

25 cm 3 – 6 m

Flexible "o" ring 10–20 cm ø 0.5–0.8 cm

Determining the pipe sizes required

12.   To select the correct size of pipes to be used on your fish farm, for example at the inlet and outlet of fish ponds, you should first know which water discharge is required in each case (see **Water**, **4**). You should then determine which size of pipe will have the capacity for such water discharge. Finally, it is best to standardize the pipes and to select only a limited number of different sizes.

13.   The water discharge capacity of a pipe increases with the pressure head (measured in cm) at the entrance of the pipe (see Section 37, **Water**, **4**). This is also shown in **Table 12** for pipes of different sizes.

TABLE 12

Water discharge capacity of concrete pipes under various pressure heads (l/s)

Inside diameter of pipe (cm)	Pressure head (cm)						
	5	10	15	20	25	100	200
20	18.7	26.4	32.3	37.3	41.8	—	—
25	29.2	41.3	50.5	58.4	65.2	120	160
30	42.0	59.4	72.8	84.0	94.0	—	—
35	57.2	80.9	99.1	114.4	127.9	190	320

Estimating pipe capacity

14.   Very often, the **pressure head varies**, for example on the outlet pipe as the fish pond is being drained. It is best, therefore, to estimate the capacity of the pipes by one of the following simple methods.

(a)   Using **Table 13** and **Graph 1**, you can estimate the water discharge capacity of pond outlet pipes of various diameters.

(b)   Using **Table 14**, you can estimate the pipe sizes required to drain a pond of a particular size in a specified time.

(c)   You can use **mathematical formulae** to estimate:

- the water **discharge capacity Q** (in litres per second, l/s) for a given pipe with an **inside diameter D** (in cm), using the formula:

$$Q = 0.078\ D^2$$

so, for a pipe with **D** = 20 cm, **Q** = 0.078 × 20^2 = **31.2 l/s**;

- the **inside diameter D** (in cm) of the pipe required for a given water discharge capacity **Q** (in l/s), using the formula:

$$D = 3.56 \sqrt{Q}$$

so, for **Q** = 16 l/s, you require a pipe with **D** = 3.56$\sqrt{16}$ = 3.56 x 4 = **14.2 cm**; and you will probably use a 15-cm pipe.

Note: all of these methods assume you are using a simple, short pipe with no obstructions to water flow, such as complicated sluice gate assemblies, screens, dirt or fouling inside the pipe's internal edges, or lips or edges at the mouth or joints of the pipe. Any of these will reduce the flow. If you have or expect to have any obstructions in the pipe, use a larger size. If the pipe is made up of several sections of different diameters, estimate the water flow on the basis of the smallest diameter pipe you use.

TABLE 13

Approximate water discharge capacity of outlet pipes*

Inside diameter of pipe (cm)	Water capacity			
	(l/s)	(l/min)	(m³/h)	(m³/24 h)
5	1.8	108	6.5	155
10	8	480	29	691
15	18	1 080	65	1 555
20	31	1 860	112	2 678
30	70	4 200	252	6 048
40	126	7 560	454	10 886
50	196	11 760	706	16 934
	x	60x	3.6x	86.4x

* Based on a pressure head of about 15 cm

TABLE 14

Time taken to drain ponds (in hours) with different drain pipe sizes

Inside diameter of pipe (cm)	Pond area (ha)					
	0.1	0.2	0.5	1	2	5
10	96	192	480	—	—	—
20	15	30	75	150	300	—
50	1.5	3.5	8	16	32	80
100	—	—	2	3.5	7	17.5

Note: these figures assume an initial internal water depth of 1 m, with pipe velocity limited to 1 m/s; for two pipes, etc., time is divided by 2

GRAPH 1

Approximate water discharge capacities of outlet pipes (pressure head about 15 cm)

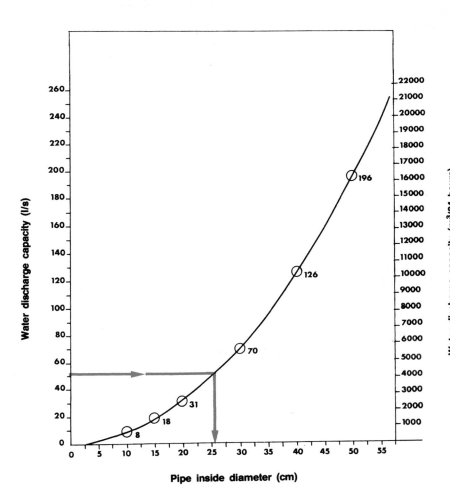

155

Designing longer pipelines

15. To design **a pipeline**, you have to use a different method to determine its water discharge capacity, taking into account its **length** and the **head loss*** from start to end. In addition, you should check that the water velocity in the pipeline will not exceed a critical value. Proceed as follows.

(a) Select a pipeline inside diameter and calculate its **water discharge capacity** (in l/s) as:

$$Q = K\sqrt{H \div L}$$

where **K** is the conveyance factor (in l/s), see **Table 15**;
H is the head loss (in m) over the pipeline length;
L is the total length (in m) of the pipeline.

Example

A concrete pipeline has an inside diameter of 20 cm. It is 100 m long (**L**), and its total head loss (**H**) is 2 m. Its water discharge capacity is:

$$Q = 399.7 \text{ l/s} \sqrt{(2 \div 100)} = 399.7 \sqrt{0.02} = 56.53 \text{ l/s}$$

A 100-m concrete pipeline with a straight run

(b) Calculate the **water velocity V** (in metres per second, m/s) in the given pipeline as:

$$V = M\sqrt{(H \div L)}$$

where **M** is the velocity modulus (in m/s), see **Table 15**;
H is the head loss (in m) over the pipeline length;
L is the total length (in m) of the pipeline.

Example

Using the same concrete pipeline with an inside diameter of 20 cm, a length (**L**) of 100 m and a total head loss (**H**) of 2 m, the water velocity is:

$$V = 12.729 \text{ m/s} \sqrt{0.02} = 12.729 \times 0.141 = 1.79 \text{ m/s}$$

(c) Compare the calculated water velocity **V** (in m/s) with the corresponding maximum velocity recommended in the last column of **Table 15**.

Example

Using the same example, the calculated water velocity **V** = 1.79 m/s exceeds the maximum recommended velocity **V** max = 0.90 m/s. Total head loss should be reduced.

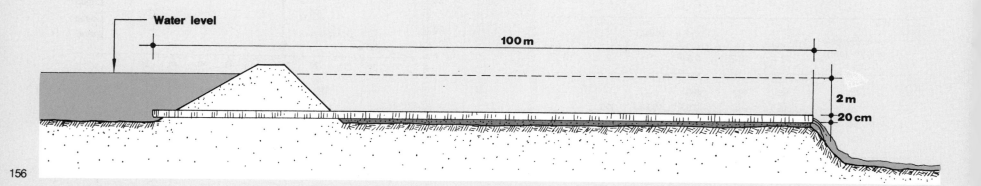

Water level

100 m

2 m

20 cm

156

TABLE 15

Basic conveyance factors for the design of pipelines

Pipe inside diameter (cm)	Plastic and new cast iron pipes		Concrete and old cast iron pipes		Sewage pipes		Maximum water velocity (m/s)
	M (m/s)	K (l/s)	M (m/s)	K (l/s)	M (m/s)	K (l/s)	
5.0	6.405	12.554	5.174	10.142	4.056	7.950	0.60
7.5	8.288	36.47	6.779	29.83	5.407	23.79	0.70
10.0	9.883	77.58	8.148	63.96	6.568	51.56	0.75
12.5	11.413	139.24	9.462	115.44	7.688	93.79	0.75
15.0	12.684	224.51	10.562	186.95	8.632	152.78	0.80
17.5	13.996	335.9	11.696	280.7	9.604	230.5	0.85
20.0	15.188	476.9	12.729	399.7	10.494	329.5	0.90
22.5	16.322	648.0	13.715	544.5	11.345	450.4	0.95
25.0	17.361	852.4	14.619	717.8	12.126	595.4	1.00
30.0	19.432	1 373.9	16.427	1 161.4	13.690	967.9	1.10

Note: M and K are constants

16. The formulae you have just used are suitable for straight pipes, but the flow of water is reduced by bends on the pipe or any fittings. The simplest way to allow for these is to think of each bend or fitting as being equivalent to an extra length of pipe of an **equivalent length**. **Table 16** shows **equivalent lengths** for typical fittings.

Example

If the pipe used previously (20-cm diameter and 100 m long) has four 90° bends, two check valves (fully open) and a reducer outlet, its discharge capacity is still:

$$Q = 399.7 \sqrt{H \div L}$$

L is now the total equivalent length (**TEL**) or the length of the pipe plus the equivalent lengths of the fittings.

Thus **TEL** = 100 m + equivalent lengths (in m) of four 90° bends + 2 check valves + reducer outlet = 100 m + 4(0.4**D**) + 2(0.75**D**) + (0.08**D**).

For the pipe diameter **D** = 20 cm, **TEL** = 100 m + 4(0.4 x 20) m + 2(0.75 x 20) m + (0.08 x 20) m = **163.6 m**.

Then **Q** = 399.7 $\sqrt{2 \div 163.6}$ = 44.19 l/s, which is less than 80 percent of the flow of the straight pipe, as calculated in the previous example.

TABLE 16

Equivalent length of pipe bends and fittings

Ref. no.	Pipe bend/fitting	Equivalent length (m)
1	Globe valve, open	2.5 D*
2	Gate valve, open	0.05 D
3	Check valve, open	0.75 D
4	45° bend	0.15 D
5	90° bend or T	0.2-0.4 D
6	Square elbow	0.7 D
7	Reducer outlet (three-quarters of original diameter)	0.08 D

Note: these are typical values and may vary according to design and manufacture
* D = pipe inside diameter (cm)

A 100-m concrete pipeline with four 90° bends

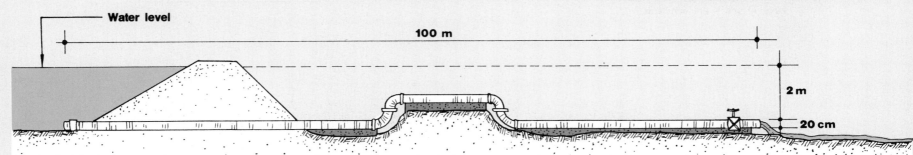

Water level

100 m

2 m

20 cm

39 Selecting a water pump

1. If you intend to use a pump, you will need to know the right size or power **P** (in kW) of the pump for the job. You need to consider the head **H** (in m), the water flow **Q** (in m³/s) and the efficiency **E** (in percent) of the pump. You can use a simple equation such as:

$$P \ (kW) = (9.81 \times Q \times H) \div E$$

where the pumping head **H** (m) is calculated as the sum of the suction head (h_s), delivery head (h_d) and pipe loss head (h_p).

(a) For the commonly used field pumps, the **suction head** (h_s) should be kept as small as possible. Most suction heads will not draw up more than 3 to 5 m in field use.

(b) The **delivery head** (h_d) is generally in the range of 2 to 10 m.

(c) The **pipe loss head** (h_p) can be calculated from the formula used in Section 38, $Q = K \sqrt{h_p \div L}$, and therefore:

$$h_p = LQ^2 \div K^2$$

where **Q** is the known water discharge (in l/s);

 L (or **TEL**) is the total length (or total equivalent length) of the pipeline (in m);

 K is the conveyance factor (in l/s), see **Table 15**;

 h_p is the pipe loss head (in m).

The definition of suction head (h_s) and delivery head (h_d)

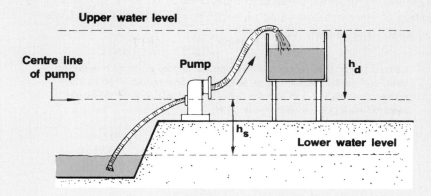

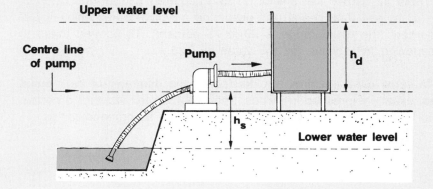

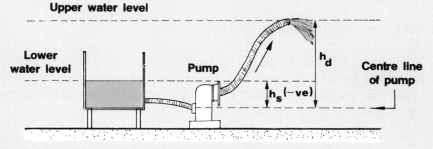

Note: h_s = distance from lower water level to pump centre line
 h_d = distance from pump centre line to upper water level

159

2. With simple, short lengths of pipe of at least the same size as the pump inlets and outlets, the pipe loss head can be ignored.

3. **Table 17** shows the power (in kW) required for various rates of flow (m³/s) and total heads (m), assuming a typical pump efficiency of 60 percent (the usual range is 40 to 75 percent). To convert these to horsepower (**HP**), divide the kW value by 0.75.

4. In some cases, pumps are defined by the **diameter of their outlet pipes**, usually expressed in inches. Then you can tell whether a certain pump is sufficient for your needs, by estimating its horsepower as

$$HP = 3.14\, D^2 \div 20$$

where **D** is the inside diameter of the outlet pipes in inches.

Note: one inch is equal to 2.54 cm.

5. If the pump is to be run for long periods of time, you should increase the power by at least 30 percent, as most pumps should not be run fully loaded for too long. The power of the motor should be at least 10 percent more than the pump power.

6. In many cases, you can select a pump using information provided by manufacturers or dealers. This is often provided as a **head-output curve** (see **Graph 2**), which shows the pumping ability of each type of pump.

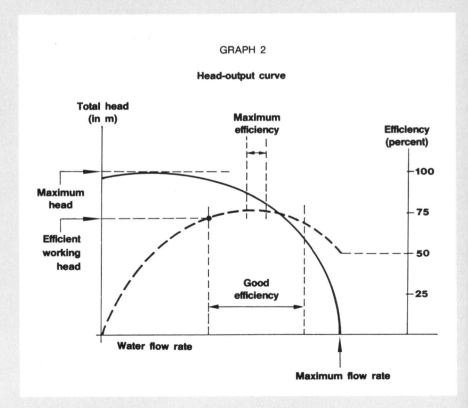

GRAPH 2

Head-output curve

TABLE 17

Pump power for different flow and head requirements*
(in kW, assuming a pump efficiency of 60 percent)

Total head (m)		Flow rate required								
	l/min: m³/s: m³/h:	10 0.00017 0.60	20 0.00033 1.20	50 0.00083 3	100 0.00167 6	200 0.00333 12	500 0.00833 30	1 000 0.01667 60	2 000 0.03333 120	5 000 0.08333 300
1		0.003	0.005	0.014	0.027	0.055	0.136	0.273	0.545	1.363
2		0.005	0.011	0.027	0.055	0.109	0.273	0.545	1.090	2.725
5		0.014	0.027	0.068	0.136	0.273	0.681	1.363	2.725	6.813
10		0.027	0.055	0.136	0.273	0.545	1.363	2.725	5.450	13.625
20		0.055	0.109	0.273	0.545	1.090	2.725	5.450	10.900	27.250
50		0.136	0.273	0.681	1.363	2.725	6.813	13.625	27.250	68.125
100		0.273	0.545	1.363	2.725	5.450	13.625	27.250	54.500	136.250
200		0.545	1.090	2.725	5.450	10.900	27.250	54.500	109.000	272.500

* To convert these kW values to horsepower, divide them by 0.75

7. If you have a choice, try to use the pump which is **the most efficient for the work**, as this reduces running costs. Efficiency is often shown on the same head-output curve or can be estimated. The pump usually works most efficiently at around 60 to 70 percent of its maximum head or output.

8. Most general-purpose field water pumps are suitable for fish farms, although if the water is brackish or carries a lot of mud, you should check that the pump is suitable. You should fit a screen at the pump intake. For centrifugal pumps (the most common type), a foot valve is useful for keeping water in the pipe when the pump is stopped. The pipe is filled with water (primed) before starting, as the pump cannot suck the water into the pipe by itself.

Several ways to start centrifugal pumps

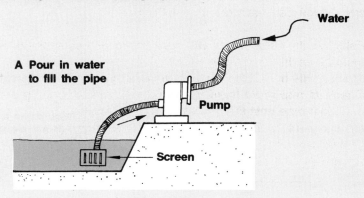

A Pour in water to fill the pipe

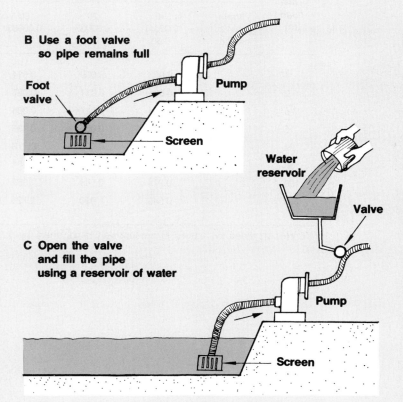

B Use a foot valve so pipe remains full

C Open the valve and fill the pipe using a reservoir of water

162

9. If a pump is already available, and you do not know its discharge capacity **Q**, you can estimate it as follows.

(a) Estimate its horsepower **HP** from the outlet pipe diameter (**D** in inches), as **HP** $= 3.14\,D^2 \div 20$.

(b) Multiply **HP** by 0.75 to obtain the pump power **P** in kilowatts.

(c) Check its maximum head **H** (in m) by running the pump and lifting the outlet pipe until the flow stops. The pump can usually operate in the range of 30 to 70 percent of this maximum head.

(d) Estimate the discharge capacity **Q** (in m³/s) from values of power (**P**) and head (**H**), as

$$\boxed{Q = (PE) \div (9.81\ H)}$$

where **E** is the pump efficiency in percent.

Example

If a pump has an outlet pipe diameter of 3 inches (7.5 cm):

$\quad$ **HP** is approximately $= 3.14\,D^2 \div 20 = $ **1.4 HP**.
$\quad$ Pump power, **kW** $= 1.4 \times 0.75 = $ **1.1 kW**.

If maximum head $= 8$ m, efficient working head is usually from 30 to 70 percent, i.e. about 2.5 to 5.5 m.

Discharge capacity, for example at 4 m, assuming a 70 percent efficiency, is **Q** $= (PE) \div (9.81\ H) = (1.1 \times 0.7) \div (9.81 \times 4) = 0.77 \div 39.24 = 0.02$ m³/s $= $ **20 l/s**.

10. You can also check the discharge capacity of the pump **Q** (m³/s) by measuring how long it takes to drain or fill a known volume of water. By estimating the total head, you can work out the pump power.

Example

If a pump fills a 50-litre barrel in 10 seconds, with an estimated total head of 10 m, efficiency is estimated at 30 percent as the pump is near its maximum head (found to be 12 m). **Q** (m³/s) $= $ volume/time $= 0.05$ m³ $\div 10$ s $= $ **0.005 m³/s**.

4 EARTHMOVING METHODS

Useful containers for earthmoving

Woven baskets

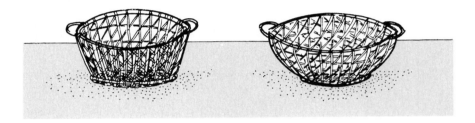

Wooden crates

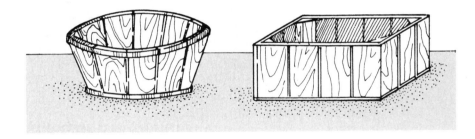

Metal containers

40 Introduction

1. Pond construction always involves the transport of earth. There are several methods available, and the choice of a particular method depends mainly on:

- local conditions such as soil on site and transport roads and their suitability for use;
- seasonal conditions;
- average transport distance;
- quantities of earth to be moved;
- work force availability and cost.

41 How to move earth by hand or using simple containers

1. The digging-and-throwing method is the simplest method.

(a) For **cohesive soils**, you can dig and cut blocks of earth. These blocks can be thrown by hand from one person to the next, along a human chain.

(b) For **all types of soil**, you can dig and throw the earth directly. The throwing distance is in practice **limited to 3 m**. (For average working standards see Section 127, **Pond construction, 20/2**).

2. If the work is being done by a fairly large group, different types of small containers can be used for earthmoving such as **woven baskets, wooden crates, metal buckets or pans**. Generally, the container is filled at the digging site and lifted by two workers who place it on the head of each member of the transporting crew.

Earthmoving by hand

**Throwing distance
is usually
limited to 3 m**

Earthmoving using containers

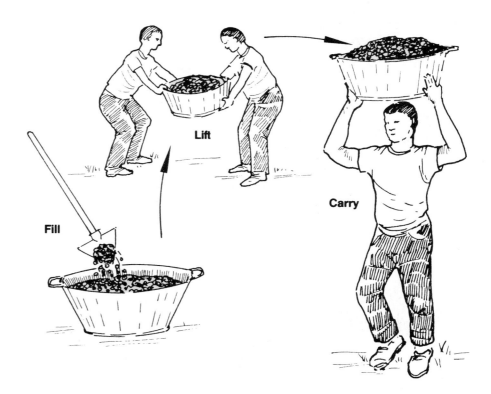

Fill

Lift

Carry

Shoulder poles

3.   Alternatively, using wooden or bamboo **shoulder poles**, each person may carry two containers balanced either on one shoulder or across two. For distances up to 50 m, the total weight should not exceed the weight of the carrier.

4. A **shoulder yoke** can be used to carry containers. The diagram on this page shows you how to carve a yoke out of a single piece of wood. The neck of the yoke should be formed so as to fit the neck of the person carrying the load.

Shoulder yokes

How to carve a shoulder yoke

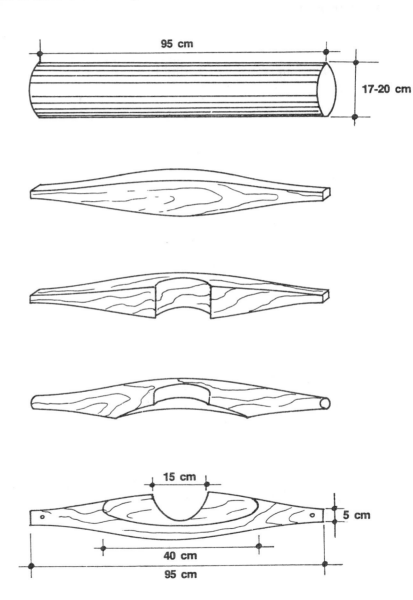

5. You can also build a **wooden transport box**, which can easily be carried by two workers. Do not build it too large or too heavy (150 × 60 × 40 cm is a good size). It is best to use light wood and to drill many holes in the bottom of the box to let any extra water run out before transport. The handles should be strong. If you use removable ones, you do not need to build a pair of handles on each transport box, but can use the same pair for several boxes.

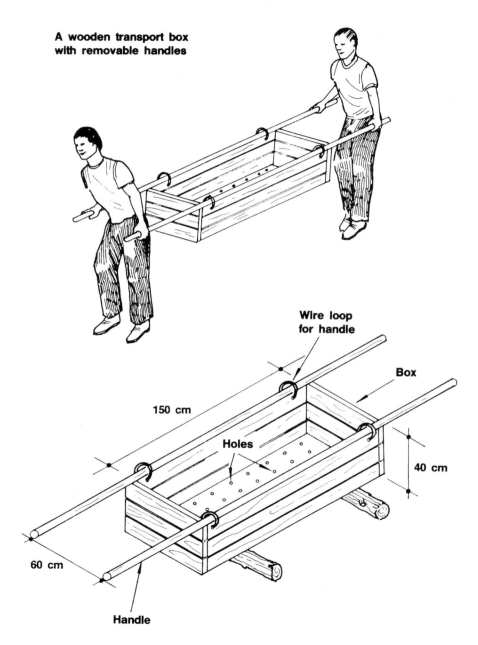

A wooden transport box with removable handles

150 cm

Wire loop for handle

Box

Holes

40 cm

60 cm

Handle

170

42 How to move earth with wheelbarrows

1.   It is usually much more efficient to move earth using wheelbarrows, but this method requires a great deal of work. In soft soils and particularly if the wheelbarrows are not equipped with a rubber tyre, walkways made of planks might be required when pushing loaded wheelbarrows. In such cases:

- use planks 4 to 6 m long, 3 cm thick and 30 cm wide;
- the transport distance is usually limited by the availability of planks to 30 m maximum.

2.   In **wet conditions** it may be difficult to use barrows on sloping ground.

3.   Standard metal wheelbarrows can transport from 40 l (0.04 m³) to 60 l (0.06 m³) of earth.

Using a plank walkway

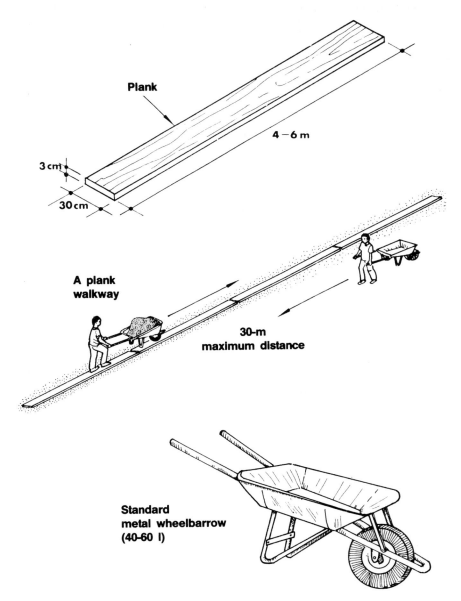

Plank

4 – 6 m

3 cm

30 cm

A plank walkway

30-m maximum distance

Standard metal wheelbarrow (40-60 l)

4. The illustrations on this page and on page 173 show designs for a number of wheelbarrows which can be made using locally available material.

The Chinese wheelbarrow

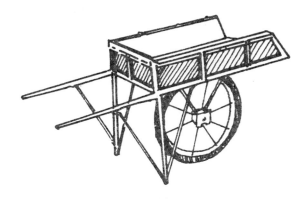

5. The **traditional northern Chinese wheelbarrow**, made with a hardwood frame and a large-diameter central spoked wheel, supports either a container above the wheel or a basket on either side of the wheel (total carrying capacity 100 to 120 kg or more). Two handles are used for pushing and balancing the barrow, which must be kept vertical while in use, as the heavy loads carried can easily tip. In addition, a strap which passes around the user's shoulders (not shown in this drawing) helps in pulling the barrow. For slopes, one or more helpers may be needed to pull the barrow. This design is very efficient for heavier loads in moderate to good ground conditions. It can be used to carry heavy loads over long distances.

Note: when this wheelbarrow is balanced near the centre of the load, very little force is needed to lift the handles.

How to make a Chinese wheelbarrow

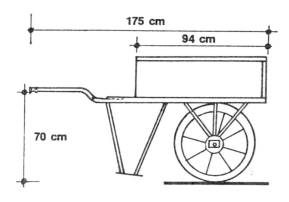

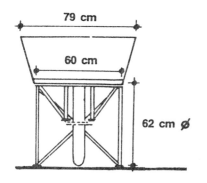

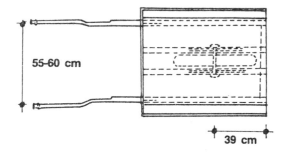

6. Other wheelbarrows that can be built using scrap wood or an old oil drum and pipe sections are also illustrated.

How to make an oil drum wheelbarrow

Wooden pole wheelbarrow

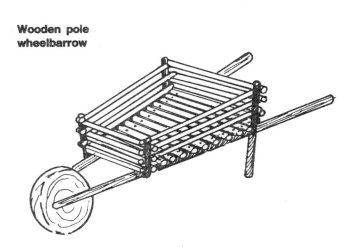

Wooden board wheelbarrow (Zaire)

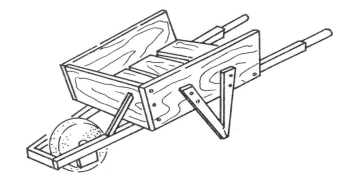

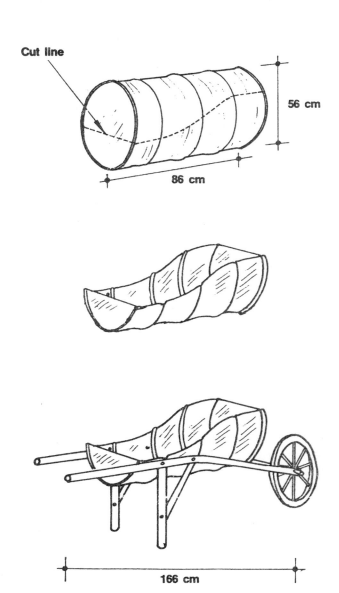

Using a wheelbarrow

7. You can plan the work outputs with a wheelbarrow by considering the following conditions.

(a) For **distances shorter than 20 m** along level ground, the total distance covered back and forth by one worker with a standard wheelbarrow averages 2 800 m per hour. For sloping ground this distance will be reduced by approximately 10 percent for a 1:50 slope and 20 to 25 percent for a 1:20 slope carrying the full load uphill. Downhill slopes will increase distances by similar amounts.

Example

The average transport distance is 17 m. Each wheelbarrow will make 2 800 m ÷ (17 m × 2) = 82 trips per working hour.

For a five-hour working day it will make 82 x 5 = 410 trips

If each wheelbarrow contains 50 l = 0.05 m^3 earth, it will transport 410 × 0.05 m^3 = 20.5 m^3 per day.

(b) To transport material for **distances up to 30 m**, for each wheelbarrow you will need at least:

● one worker to dig and fill the wheelbarrow;
● one worker to push the wheelbarrow.

(c) You might need additional workers at particular spots:

● along the transport road at climbing points, to help bring the wheelbarrow up the slope;
● at the dumping site, to help empty the wheelbarrow completely.

Transporting earth by wheelbarrow

One worker to push

One worker to dig and fill

Help at climbing points

Help at dumping points

43 Draught animal power

1. Draught animal power has been used for thousands of years to provide power for traction, for example for pulling carts, scrapers and scoops. The most commonly used animals are oxen, buffaloes and donkeys.

2. For any species, provided the animals are in good condition, the power and pull capability depends mainly on the weight of the animal and on the ground conditions, as shown in the following chart.

Animal	Typical weight (kg)			Pull:weight ratio*		Typical speed (km/h)
	Light	Average	Heavy	Low	High	
Ox	250	350-600	700	0.10	0.12	3.0
Cow	220	300-500	600	0.08	0.10	2.5
Buffalo	400	500-700	800	0.12	0.14	3.0
Donkey	120	150-200	250	0.14	0.20	3.5

* The pull:weight ratio is based on the continuous capability of the animal over a working period of three to four hours. The lower value of pull:weight ratio should be used when walking conditions are poor (i.e. rough or soft ground); the higher value may be used when conditions are good

Example

A 450-kg ox walking over good ground has a pull:weight ratio of 0.12. It should be capable of providing a pull of about 0.12 x 450 kg = 54 kg

3. For pulling **carts with wooden wheels** on damp and flat agricultural soil, you should check that the pull capability of your animal equals at least 10 percent of the total load, that is cart weight plus earth load. If the ground is sloping, you should increase this percentage accordingly. If you use pneumatic tyred wheels, you might reduce the pull capability to 5 to 8 percent of the total load.

Example

If your cart with wooden wheels weighs 250 kg and its useful load of earth is 1 000 kg, you will require a minimum pulling capability of (1 000 kg + 250 kg) × 0.10 = 1 250 kg × 0.10 = 125 kg to pull it over damp or flat agricultural soil. On a 5-percent slope, you should add (125 kg x 0.05) = 6.25 kg, totalling 131.25 kg. If your cart has pneumatic tyred wheels, on a 5-percent slope you will require a pull capability of about 131.25 kg × 0.70 = about 92 kg.

4. There are several types of **harness** in use. The most effective type depends on the species and breed of animal:

- **yokes** are best for bovines, which have strong shoulders;
- **collars or breastbands** are best for donkeys, horses and mules, which produce their best pull from the breast.

5. **Yokes** may be used for single animals or to harness animals together in pairs, but they should be adapted to the species of bovine used, for example:

- with *zebus* use the shoulder yoke;
- with *n'damas* use the neck yoke or forehead yoke.

Note: various kinds of animal harness are shown on page 176 and two kinds of yoke that you can easily make are shown on page 177.

Various kinds of animal harness

Double head yoke

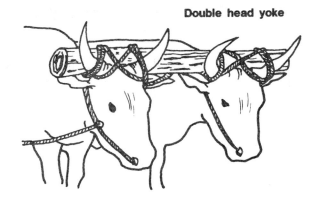

Double shoulder yoke

Single shoulder yoke

Breastband

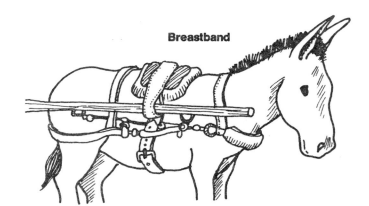

Three-pad collar

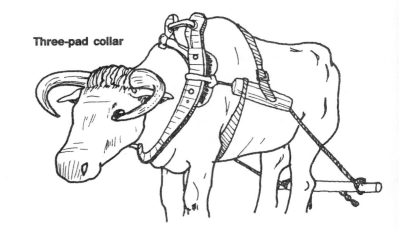

176

How to make a double shoulder yoke

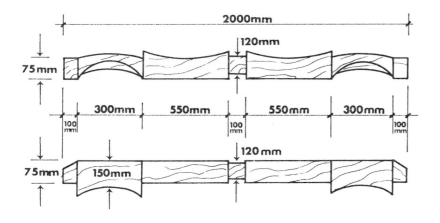

How to make a double head yoke

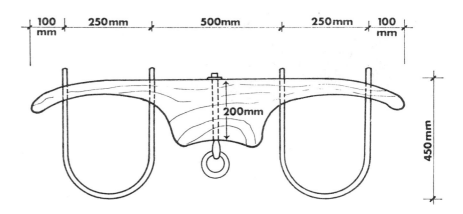

Lash the yoke to the horns

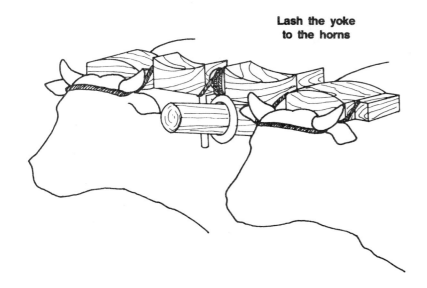

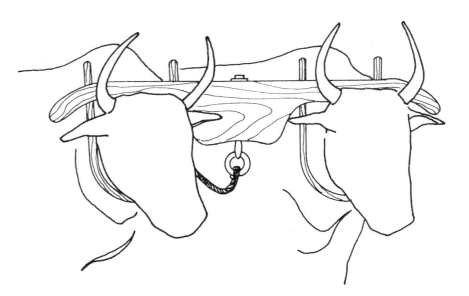

1. To move earth over longer distances, simple carts can be either pushed or pulled by hand or drawn by animals. Soil conditions must be favourable. Transport will be easier if a track or road with a hard, clean surface is used. Remember that a well-made cart must be correctly balanced when empty. Also keep it balanced when loaded, and keep the wheel hubs and axles well greased.

2. Carts can be built very simply using wood or bamboo. Wooden wheels can be used, or old car wheels can be fitted by most local workshops. This will improve the handling of the cart and make it easier to pull. If the load is properly balanced, a load of up to 400 kg for a donkey and 800 kg for a pair of oxen is manageable on flat and firm ground (see Section 43). A 1:20 uphill slope reduces this by 50 to 60 percent; a 1:50 slope reduces the load by about 20 percent.

Various carts suitable for moving earth

Zebu or ox cart

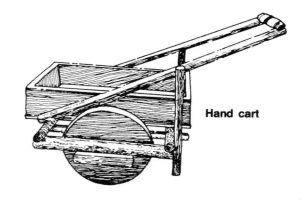

Hand cart

Donkey cart

3. You can improve the use of your cart for carrying earth by placing on its floor two **fibre mats with strong handles** on each outer side. Load the earth on top of these mats. When the cart is full, bring the sides of the mats up and firmly tie the two handles together. To unload, detach the ropes keeping the sides of the mats vertical and pull the package of mats off the cart floor. Move the cart forward and free the mats from under the dumped earth.

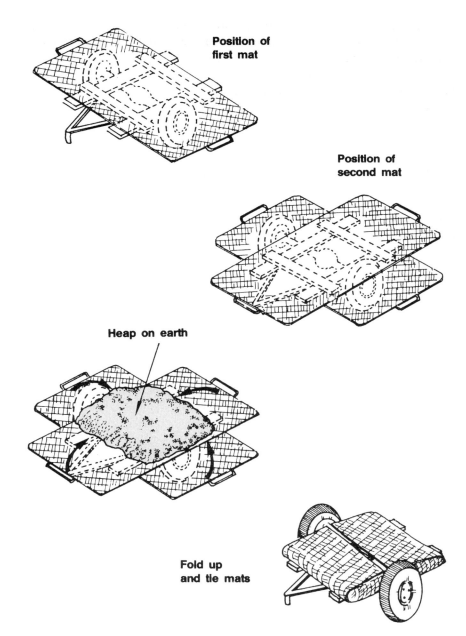

Position of first mat

Position of second mat

Heap on earth

Fold up and tie mats

Using fibre mats and a two-wheel cart to carry earth

A two-wheeled cart

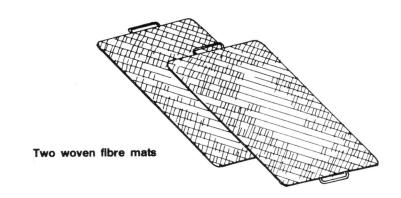

Two woven fibre mats

4. You can also use a **tipping cart** to carry earth. Notice that a chain at the front of the cart can be released when you are ready to tip the cart and dump the load.

**A tipping cart
for moving earth**

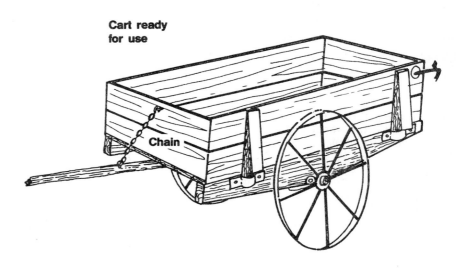

Cart ready
for use

Chain

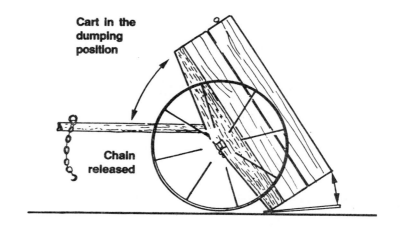

Cart in the
dumping
position

Chain
released

45   How to move earth with an oil drum scraper

1.   A scraper is a piece of equipment, usually animal drawn, which fills with earth as it is moving. When full, the earth is transported over a short distance and dumped. This device is particularly useful for levelling off uneven ground.

2.   The scraper can be built by a blacksmith from an old oil drum and scrap metal. It is cheap to build and moves earth quite efficiently over short distances (70 to 200 m) when pulled, for example, by a team of oxen and operated by one person.

3.   It works best when the soil has been loosened up first, by ploughing for example. It is estimated that under normal use the oil drum scraper will last for at least five years.

Materials for an oil drum scraper

- 1 oil drum (200 l), strong and without rust;
- 1 metal blade, 5 to 8 mm thick, 88 cm long, tapered to a sharp edge along one side (old truck springs make good blades);
- 2 metal blade holders, 5 to 8 mm thick;
- 1 softwood handle, 3 m long, 4 × 8 cm or an 8-cm diameter pole;
- 1 wood handle brace, 150 cm long, 3 × 8 cm;
- 1 wood block, 12 cm long, 3 × 8 cm;
- 1 steel bolt, 1 cm in diameter, 10 cm long;
- 5 nails, 9 cm long;
- wire, at least 3 mm thick, 12 m long;
- chain, links 7 mm thick, 4 m long, with hook at each end;
- rope, 12 mm in diameter, 3 m long.

Note: you will be shown how to make an oil drum scraper on pages 182 to 184 and how to use a scraper on pages 185 and 186.

Using an oil drum scraper

Parts of an oil drum scraper

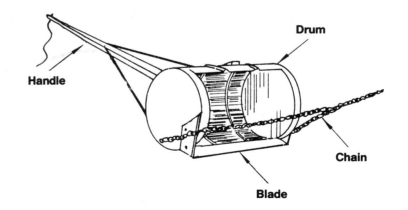

Handle

Drum

Chain

Blade

4. **Cut the side** of the drum halfway around, starting next to the welded seam and **leaving 6 cm of metal** on each side to form runners, except for the last 30 cm where it should be cut next to the edges of the drum.

5. **Pull the cut-out section forward** and flatten it with a hammer. Fold the cut-out section back 17 to 20 cm from the end of the cut to **form a double bottom**. **Weld** it in place.

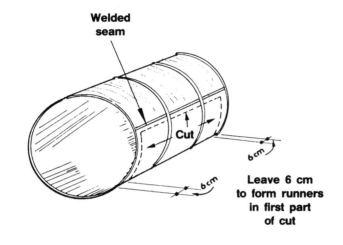

Welded seam

Cut

6 cm

Leave 6 cm to form runners in first part of cut

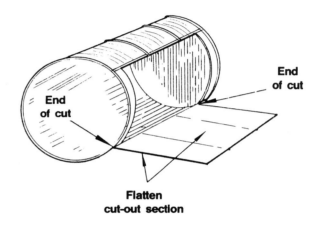

End of cut

End of cut

Flatten cut-out section

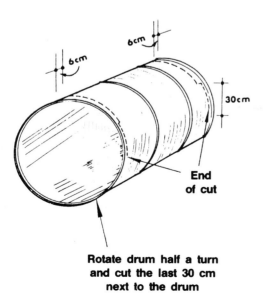

6 cm

6 cm

30 cm

End of cut

Rotate drum half a turn and cut the last 30 cm next to the drum

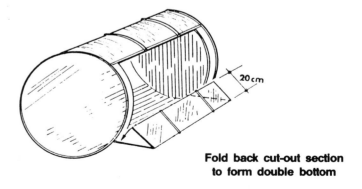

20 cm

Fold back cut-out section to form double bottom

6. **Butt the blade** against the foot of the drum fold, with its sharpened edge directed forward. **Weld** it with five spots of welding 3 cm long, evenly spaced.

7. On each side of the blade, **place a metal blade holder vertically**, the lower tip being level with the end of the cut. **Weld** the blade holders to the outside of the drum and to its heavy outer rims. **Weld** the blade to the inside of each blade holder.

8. **Taper the end of the handle** for its last 30 cm. Angle the other end to fit the drum.

9. Position the drum so that the edge of the blade is **exactly 4 cm above the ground**. Place the handle on the ground, at the centre of the bottom of the drum and at a right angle to its axis.

10. **Punch a hole** through the bottom of the drum. Drill a hole through the end of the handle. Bolt the handle to the drum.

11. **Bend** 2.5 cm of the upper edge of the drum metal up. **Punch two small holes** in the metal and drive two nails through the holes into the end face of the wooden handle brace.

12. Check that the blade is still 4 cm above the ground and **nail** the wooden block on to the handle, against the lower end of the brace. **Drive a nail** through the end of the brace into the handle.

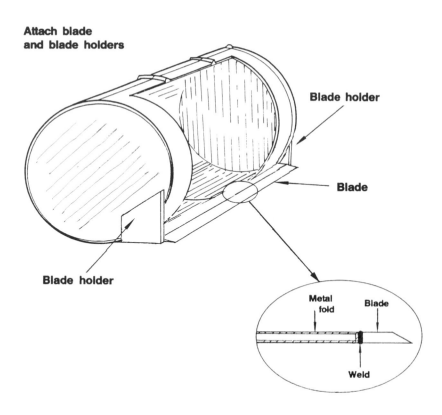

Attach blade and blade holders

Blade holder

Blade

Blade holder

Metal fold · Blade

Weld

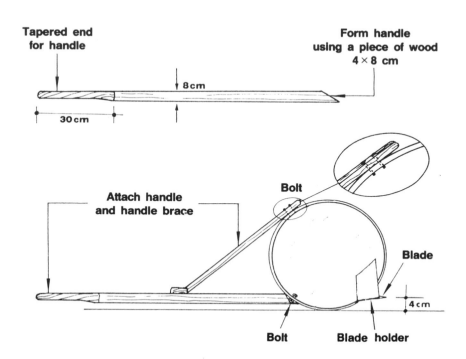

Tapered end for handle

Form handle using a piece of wood 4 × 8 cm

8 cm

30 cm

Attach handle and handle brace

Bolt

Blade

Bolt Blade holder

4 cm

13. **Drill a hole** 1.5 cm in diameter and 20 cm from the end of the handle to be used to attach a piece of rope. **Drill a hole** through both the handle brace and the handle to bolt them together.

14. **Punch holes** through each end of the drum for the traction chain as shown below. **Punch holes** through each end of the drum to attach wire braces to stabilize the handle as shown below.

15. **Thread the end of the rope** through the handle and knot it securely in place.

16. **Thread four strands of wire** through the holes at the back of the drum and secure them to the handle brace. If necessary, twist the wires with a long nail to tighten the brace, but make sure that the handle is at right angles to the drum.

17. **Attach the hooks of the traction chain** to the holes at the front of the drum.

Drill and punch holes

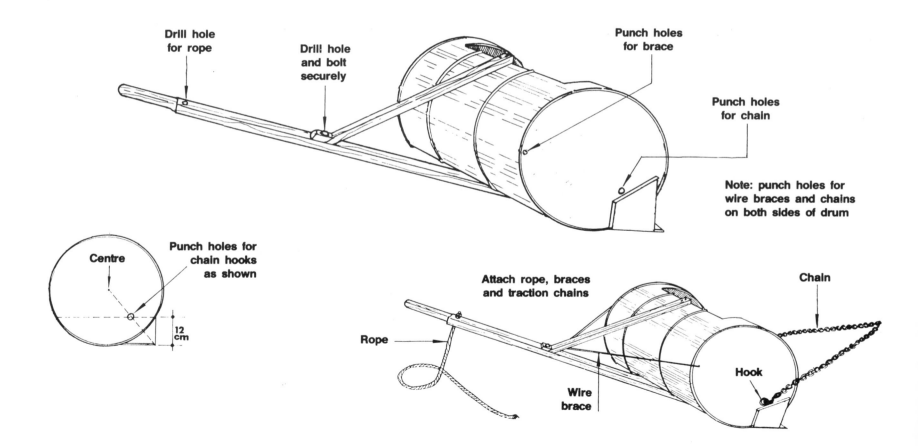

Drill hole for rope

Drill hole and bolt securely

Punch holes for brace

Punch holes for chain

Note: punch holes for wire braces and chains on both sides of drum

Centre

Punch holes for chain hooks as shown

12 cm

Attach rope, braces and traction chains

Chain

Rope

Wire brace

Hook

Operating the oil drum scraper

18. Before using the scraper, plough the ground where you want to remove soil. This will make it easier to load the soil into the scraper.

Plough ground

19. Move the scraper into position by pushing down on the handle to lift the blade off the soil. Let the animals slide the scraper forward to the point where you want to start removing soil.

Push down handle to slide scraper

20. As the scraper moves forward, start loading it by lifting the handle to let the blade dig slightly into the soil. Do not make too deep a cut, as this would either turn the scraper over or pull the animals to a stop.

Load by lifting handle

21. When the scraper is full, push down on the handle and let the scraper slide forward to where you want to unload it.

When loaded push down handle

22. To unload, lift the handle so that the animals can pull the scraper into the dumping position.

**Lift handle
to unload**

BE CAREFUL

Never have any part of your body directly above the handle of the scraper while operating it. **Always** keep a firm grip on the handle while loading or getting ready to unload, controlling it at all times.

Note: the power that is used to pull the scraper will also help in operating it. You will learn by experience how to hold the handle for a proper cut and smooth handling.

23. **Use the rope** to control the scraper's position:

- when spreading the soil evenly, hold the rope tight;

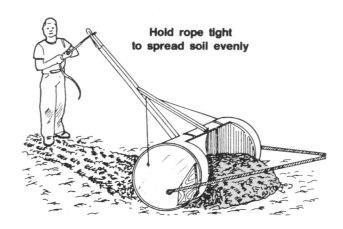

**Hold rope tight
to spread soil evenly**

- when dumping the soil in a pile, let go of the rope.

**Let go of rope
to dump soil in pile**

46 How to move earth with a dam scoop

1. The dam scoop, or soil scoop, is a simple, robust animal-drawn implement which can be either bought or made locally. It is similar to but generally smaller than the oil drum scraper described in Section 45, its capacity varying from 50 to 150 litres.

2. The dam scoop should be used on dry ground which is sufficiently loose so loading will not be too difficult. Loading is automatic, and unloading is done by tipping over where desired using a simple stop-and-catch device. The transport distance to be covered should be fairly short, not more than 40 to 50 metres.

**Two examples
of dam or soil scoop**

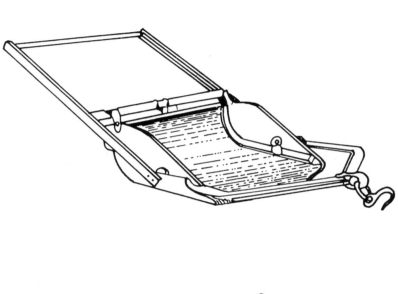

**Excavating a pond
using an ox-drawn scoop**

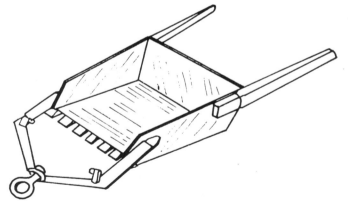

47 How to move earth with a boat, float or raft

1. In areas which are flooded (even temporarily) or adjacent to rivers or canals, simple flat-bottomed boats or floats can be useful for moving earth. In some cases it may even be worthwhile to flood all or part of a pond during construction to simplify the movement of earth, as using a float or boat requires much less work than movement by cart or scraper.

2. For simple construction work, a load of from 500 kg to 2 to 3 tonnes is common. As a rough guide:

- **a flat-bottomed boat** 5 m long by 1.5 m wide can carry about 1 000 kg of soil;
- **a wood and plywood float** 3 m long by 1 m wide and at least 30 cm deep can carry about 700 kg;
- **a raft** supported by four oil drums can carry about 1 500 kg when fully loaded.

3. If the water is shallow, it may be necessary to reduce the load to keep the boat or float high enough above the water level.

**How to build
a float or a raft**

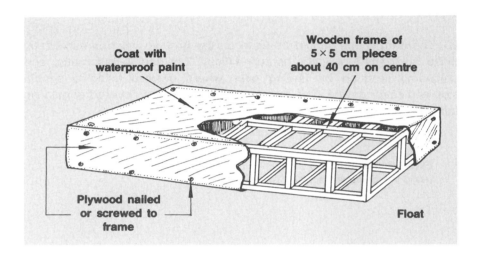

Coat with waterproof paint

Wooden frame of 5 × 5 cm pieces about 40 cm on centre

Plywood nailed or screwed to frame

Float

Flat-bottomed boat

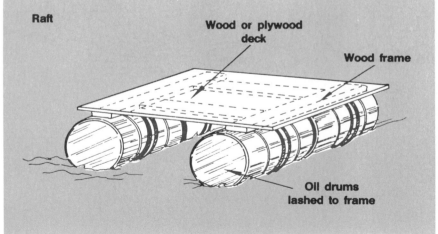

Raft

Wood or plywood deck

Wood frame

Oil drums lashed to frame

48 Earthmoving machines

1. When relatively **large areas of land or large volumes of earth** are involved, machines are generally preferred to move earth, if local conditions make it possible.

2. For maximum efficiency and minimum cost, the choice of an earth-moving machine should take into account the distance of transport, the local conditions, the period the machine can operate daily and the characteristics of the earth to be moved. Remember also that under wet soil conditions, a machine with tracks will usually be more efficient than one with wheels.

3. There are various **earthmoving activities**: loosening and digging up the earth, moving it, lifting it up to another area or into another machine, and placing it in or on a specific location. Other machines (see later) can be used for compacting the final earth construction. Some machines are built for specific tasks, others are capable of more than one task.

4. Although the selection and planning of the use of the most efficient machinery is often best left to an experienced local contractor, **Table 18** provides some guidance to use in normal field conditions.

5. The average hourly output of bulldozers and wheel-loaders for different types of earthwork are given in **Tables 19 and 20**.

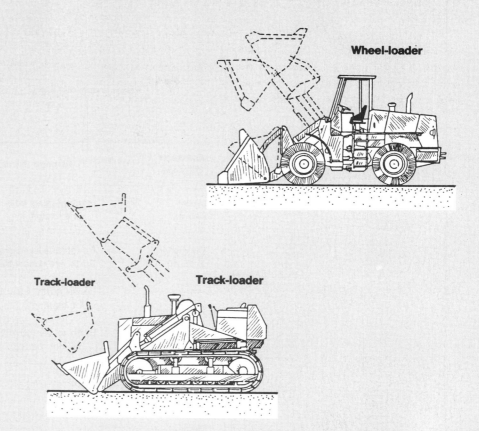

Wheel-loader

Track-loader   Track-loader

Using a track-loader

189

TABLE 18

Earthmoving machines: guidelines for possible choices

Transport distance	Working conditions	Preferred choice of machine*
Direct (0-5 m)	Placing/lifting adjacent to works, small capacity	Backhoe
Very short (5-15 m)	Rough/uneven terrain Earth not concentrated	Track-loader
	Loose earth, easy to dig for loading	Wheel-loader
	Earth to be moved, but not loaded	Bulldozer
Short (15-70 m)	—	Bulldozer
Medium (70-250 m)	—	Towed scraper
Long (250-1 000 m)	—	Motorscraper
Very long (over 1 000 m)	—	Truck

* Under normal field conditions, namely flat topography and dry weather

TABLE 19

Average output of various machines per working hour

Machinery and use	Unit	Haul distance		
		0-20 m	50 m	100 m
Bulldozer				
Clearing bush, diam. 0-6 cm	m²	—	300	—
diam. 6-10 cm	m²	—	200	—
Felling trees, diam. <20 cm	piece	30	—	—
diam. 20-40 cm	piece	10	—	—
diam. >40 cm	piece	2-5	—	—
Cutting/moving surface soil				
thickness 10 cm	m²	—	400	300
thickness 11-20 cm	m²	—	200	150
Levelling,				
thickness 10 cm	m²	800	—	—
thickness 11-20 cm	m²	400	—	—
Cutting/moving firm earth	m³	40	20	10
Moving loose earth	m³	90	50	20
Building dikes	m³	100	—	—
Wheel loader				
Loading loose earth	m³	60-100	—	—
Compactors (25 cm layers)				
Sheepsfoot roller	m²	1 000	—	—
Steel-wheel roller	m²	2 000-5 000	—	—
Rubber-tyred roller	m²	5 000-15 000	—	—
Platform vibrator	m²	300-600	—	—
Frog	m²	30-150	—	—

TABLE 20

Approximate output of bulldozers for earthwork

Approx. power rating (horsepower)	Approx. blade capacity* (m^3)	Excavation/ transport** (m^3/h)	Spreading loose earth (m^3/h)
40	1.2	13-17	18-24
70	2.5	22-29	30-39
90	3.6	32-40	42-54
130	4.0	46-71	60-76

* When completely filled. In practice, filling is usually 30 to 60% of this, depending on site conditions
** Excavation by layers less than 0.5 m thick. Transport over 50 m at most, in good site conditions

Note: it is possible to estimate output by defining the time for each trip based on: excavation of blade load 0.5-1 min; pushing 2 km/h; returning 4-5 km/h; turning, positioning, gear change 0.5-1 min; allow a maximum of 50 min per hour utilization. These output figures decrease considerably in difficult site conditions such as sloping ground

6. **Loaders** either on wheels or on tracks are not only very efficient for loading material on to lorries but also for excavating thin soil layers over large areas and for clearing forested areas (see Chapter 5). Their bucket size varies from 0.5 m^3 to 6 m^3. A loader with a 0.75 m^3 bucket, for example, equals approximately the loading capacity of a team of 30 workers.

7. A **backhoe** is a powerful arm and bucket mounted at the back of a tractor and hydraulically controlled. It can be extremely useful, particularly for small works. It is a multipurpose machine, particularly efficient for digging ditches and foundations, placing material nearby and building up small dikes. Many backhoes are also equipped with a front-end bulldozer blade, although their capacity is much smaller than that of a specialized bulldozer.

Small backhoe on tracks

Small backhoe on wheel-tractor

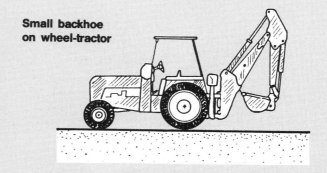

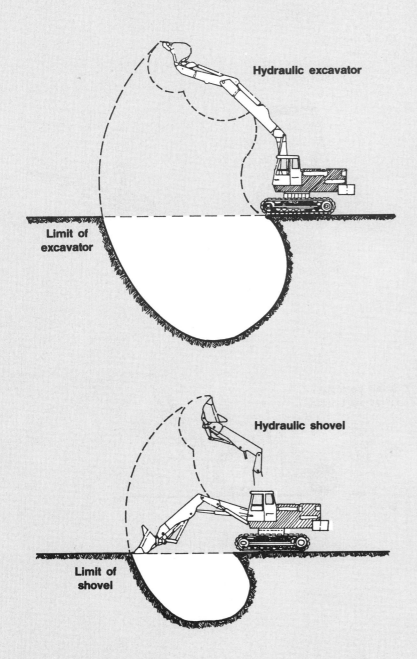

Hydraulic excavator

**Limit of
excavator**

Hydraulic shovel

**Limit of
shovel**

8. Where heavy and difficult digging is required, where it is necessary to dig under water, to finish side slopes or to dig trenches and canals, other machines may be used:

- **hydraulic excavator**, with a 360° revolving backhoe;
- **hydraulic shovel**, similar except for the front-loading bucket;
- **dragline excavator**, with a large bucket controlled by cables.

9. However, these specialized machines are expensive to operate and would only be used in difficult or unusual circumstances.

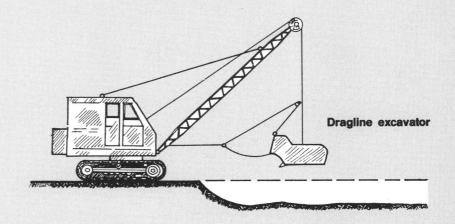

Dragline excavator

194

5   PREPARATION OF THE CONSTRUCTION SITE

1. The construction site is usually prepared in two steps: first **the vegetation is cleared**, and then **the surface soil layer is removed**. These operations can be done either by hand or by machine. In both cases, special pieces of equipment such as ropes, cables and chains may be required. You will learn about these first, to be able to make the best and safest use of them.

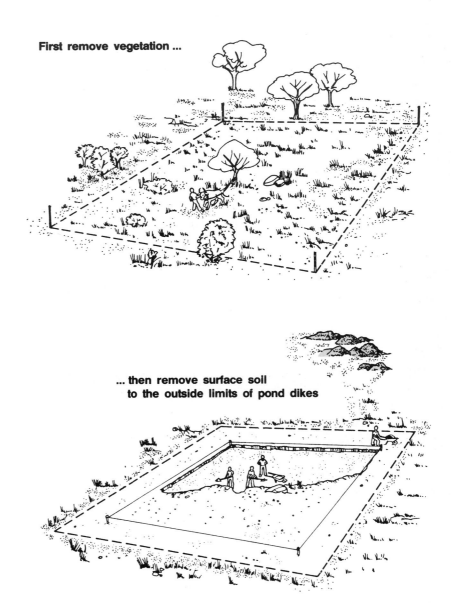

First remove vegetation ...

... then remove surface soil to the outside limits of pond dikes

Using the equipment

1. Ropes, cables, chains, pulley blocks and fittings are normally used for pulling down and clearing heavier trees, brush, etc., for pulling out rocks or other obstructions, and for moving heavy equipment around the site. The important factors are the pulling or traction load and the strength of the different components. The load can be applied either by hand, using animal power, or by machine. **Table 21** shows typical attainable **pulling loads**. If a pulley or block is being used (see pages 227 to 229), these loads may also be multiplied several times.

Pulley blocks

Rope

Cable

Chain

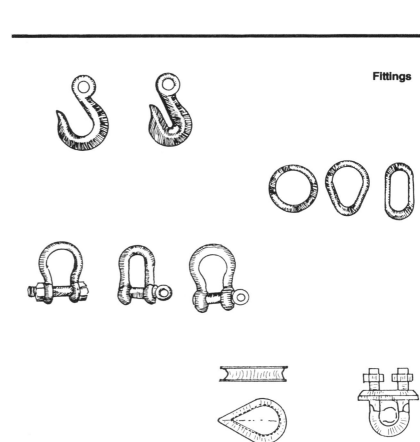

Fittings

TABLE 21

Typical attainable pulling or traction loads

Agent	Weight (kg)	Pull (kg) *
Human	50-70	10
Donkey	250	35
Mule	400	50
Cow	500	50
Bullock/ox	600	70
Horse	1 000	80
Pickup truck, 70 HP	1 500	200-300
Crawler tractor		
95 HP	8 500	3 000 - 7 500
160 HP	14 500	7 000 - 14 000
240 HP	21 500	11 000 - 20 000

* For tractors, pull depends on gear ratio; these figures are for drawbar pull in first to third gear. For humans and animals, loads quoted are for a continuous moving pull. Short "static" pull loads may be two to three times the figures given

2.   Ropes can be made of either vegetal fibres or synthetic fibres, and their characteristics vary accordingly. In particular, the strength (expressed in kg as the breaking load) is greater for synthetic ropes. The breaking load also increases as the rope becomes thicker.

3.   To determine which rope you should use, consult:

● **Table 22**, for manila, sisal or hempen ropes;
● **Table 23**, for synthetic ropes.

4.   Then find the **breaking load (BL)** for a particular diameter. To find the **safe working load (SWL)**, divide BL by the given **safety factor (SF)**, thus

$$\boxed{\text{SWL} = \text{BL} \div \text{SF}}$$

5.   **Vegetal fibre ropes** may rot or lose strength if old and poorly stored. If in doubt about the quality of the rope, you should increase the safety factor or, ideally, test the rope in safe conditions with a load at least two to three times the expected load.

6.   Similarly if any of the rope has been overloaded, i.e. near or up to its breaking load, it will be weakened for subsequent use, by 50 percent or more in some cases. Knots, splices, sharp bends, etc. will also weaken the rope.

7.   Another important factor is the elasticity of the rope. While a springy, elastic rope may be useful for giving a quick pull, if the rope springs back excessively when the load is reduced (e.g. a stump being pulled out of the ground or the rope breaking), the whiplash effect could be dangerous.

Example

(a)   **A sisal rope** of standard quality has a 24-mm diameter. Using **Table 22**, such a rope has a breaking load BL = 2 720 kg. The safety factor SF = 20. Then the safe working load SWL = 2 720 kg ÷ 20 = 136 kg.

(b)   **A polyester synthetic rope** with a 24-mm diameter has a breaking load BL = 9 140 kg (**Table 23**). The safety factor SF = 20. Then the safe working load SWL = 9 140 kg ÷ 20 = 457 kg.

TABLE 22

Characteristics of some vegetal fibre ropes*

| Diameter (mm)** | Safety factor | Hemp | | | | Manila | | | | Sisal | | | |
| | | Untreated | | Tarred | | Standard quality | | Extra quality | | Standard quality | | Extra quality | |
		Weight (kg/100 m)	Breaking load (kg)	Weight (kg/100 m)	Breaking load (kg)	Weight (kg/100 m)	Breaking load (kg)	Weight (kg/100 m)	Breaking load (kg)	Weight (kg/100 m)	Breaking load (kg)	Weight (kg/100 m)	Breaking load (kg)
10	25	6.6	631	7.8	600	6.2	619	6.2	776	6.4	487	6.4	619
11	25	8.5	745	10.0	708	9.15	924	9.25	1 159	8.4	598	9.0	924
13	25	11.3	994	13.3	944	11.2	1 027	12.4	1 470	10.9	800	11.0	1 027
14	25	14.3	1 228	17.0	1 167	14.2	1 285	15.0	1 795	12.5	915	14.0	1 285
16	25	17.2	1 449	20.3	1 376	17.5	1 550	18.5	2 125	17.0	1 100	17.2	1 550
19	25	25.3	2 017	29.8	1 916	25.5	2 230	26.65	2 970	24.5	1 630	25.3	2 230
21	20	30.0	2 318	35.4	2 202	29.7	2 520	30.5	3 330	28.1	1 760	29.0	2 390
24	20	40.2	3 091	47.4	2 936	40.5	3 425	41.6	4 780	38.3	2 720	39.5	3 425
29	20	59.0	4 250	70.0	4 037	58.4	4 800	59.9	6 380	54.5	3 370	56.0	4 640
32	15	72.8	5 175	86.0	4 916	72.0	5 670	74.0	7 450	68.0	4 050	70.0	5 510
37	15	94.8	6 456	112.0	6 133	95.3	7 670	98.0	9 770	90.0	5 220	92.0	7 480
40	10	112.0	7 536	132.0	7 159	112.5	8 600	115.8	11 120	—	—	—	—

* For dry ropes in good condition
** In some countries the thickness of ropes is expressed by their perimeter length in inches;
you can quickly calculate the approximate equivalent rope diameter (in mm) by multiplying the perimeter in inches by 8

TABLE 23

Characteristics of some synthetic fibre ropes
(hawser lay)

Diameter (mm)*	Safety factor SF	Polyamide (PA)		Polyester (PES)		Polypropylene (PP)		Polyethylene (PE)	
		Weight (kg/100 m)	Breaking load (kg)	Weight (kg/100 m)	Breaking load (kg)	Weight (kg/100 m)	Breaking load (kg)	Weight (kg/100 m)	Breaking load (kg)
8	25	4.2	1 350	5.1	1 020	3	960	3	685
10	25	6.5	2 080	8.1	1 590	4.5	1 425	4.7	1 010
12	25	9.4	3 000	11.6	2 270	6.5	2 030	6.7	1 450
14	25	12.8	4 100	15.7	3 180	9	2 790	9.1	1 950
16	25	16.6	5 300	20.5	4 060	11.5	3 500	12	2 520
18	25	21	6 700	26	5 080	14.8	4 450	15	3 020
20	20	26	8 300	32	6 350	18	5 370	18.6	3 720
22	20	31.5	10 000	38.4	7 620	22	6 500	22.5	4 500
24	20	37.5	12 000	46	9 140	26	7 600	27	5 250
26	20	44	14 000	53.7	10 700	30.5	8 900	31.5	6 130
28	20	51	15 800	63	12 200	35.5	10 100	36.5	7 080
30	15	58.5	17 800	71.9	13 700	40.5	11 500	42	8 050
32	15	66.5	20 000	82	15 700	46	12 800	47.6	9 150
36	15	84	24 800	104	19 300	58.5	16 100	60	11 400
40	10	104	30 000	128	23 900	72	19 400	74.5	14 000

* In some countries the thickness of ropes is expressed by their perimeter length in inches;
you can quickly calculate the approximate equivalent rope diameter (in mm) by multiplying the perimeter in inches by 8

Selecting cable

8. A cable is a wire rope made up of three parts:

- **the core** at the centre can be either a fibre core (sisal or manila) or a steel core (miniature cable);
- **a number of strands** are wound around the core;
- **a number of steel wires** are wound into strands; these wires can be arranged in one or more layers.

9. A cable is designated by its **number of strands** (not including the core) and by the **number of wires** in each strand. It is also important to know the size of the cable and the grade of steel wire used.

Example

A widely used cable is the 6 x 19 type: it has six strands and each strand has 19 wires. These wires can be arranged, for example, in three layers:

- 12 wires (outside) + 6 wires (middle) + 1 wire (centre);
- 9 wires (outside) + 9 wires (middle) + 1 wire (centre).

10. The characteristics of some of the most common six-strand cables, made of steel wires with BL = 140 kg/mm^2, are given in **Table 24**, for diameters ranging from 10 mm to 26 mm:

- their weight (W) in kg per 100 m;
- their breaking load (BL) in kg;
- their safe working load (SWL) in kg, based on a safety factor equal to five.

11. To determine which six-strand cable to use, find from **Table 24** the cable size required to resist the working load you plan to apply to the cable (see Section 54).

Example

You are using a three-tonne hand winch. From **Table 24**, you will require an 18-mm diameter 6 × 19 cable, type 12 + 6 + 1 with fibre core (column 2) or type 9 + 9 + 1 with steel core, to have a SWL of at least 3 000 kg.

Maintaining cable quality

12. Cables should be fitted on to other components using proper eyes or clamps and should not be knotted. A properly fitted eye will reduce the overall breaking load by 10 to 15 percent (see paragraph 20).

13. Cables should be kept well oiled or greased, laid or hung straight (short lengths) or properly coiled. Be very careful in using cable which is kinked, flattened, rusted or has broken strands. If you must use it, reduce working loads by at least 50 percent, and avoid using it where it may stick or jam (e.g. with pulley blocks). You should at all times keep clear of the cable when it is under load.

Note: to measure the diameter of a cable correctly, use calipers if possible.

**Examples of wire cable
and how it is made up**

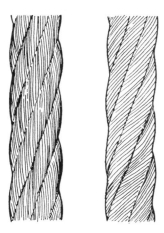

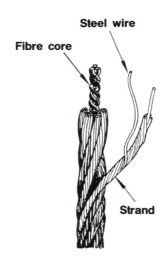

Steel wire

Fibre core

Strand

**Measuring the diameter
of cable using calipers**

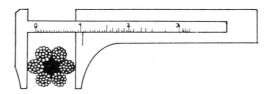

YES

Using slide calipers

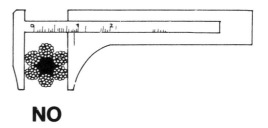

NO

Cross sections of six-strand cables

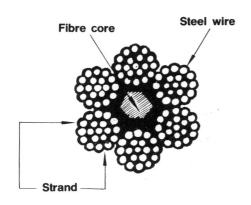

Fibre core

Steel wire

Strand

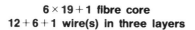

**6 × 19 + 1 fibre core
12 + 6 + 1 wire(s) in three layers**

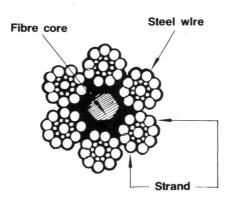

Fibre core

Steel wire

Strand

**6 × 19 + 1 fibre core
9 + 9 + 1 wire(s) in three layers**

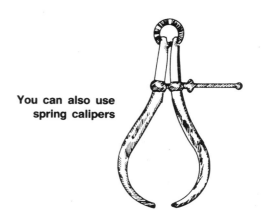

**You can also use
spring calipers**

203

TABLE 24

Characteristics of some galvanized steel cables*

Diam. (mm)	Cables with fibre core									Cables with steel core					
	6 × 12 (one layer 12 wires)			6 × 19 (3 layers 12+6+1 wires)			6 × 7 (2 layers 6+1 wires)			6 × 19 (3 layers 12+6+1 wires)			6 × 19 (3 layers 9+9+1 wires)		
	W (kg/ 100 m)	BL (kg)	SWL (kg)	W (kg/ 100 m)	BL (kg)	SWL (kg)	W (kg/ 100 m)	BL (kg)	SWL (kg)	W (kg/ 100 m)	BL (kg)	SWL (kg)	W (kg/ 100 m)	BL (kg)	SWL (kg)
10	24.3	3 020	600	41	5 400	1 080	34.7	4 820	960	33.6	4 460	890	—	—	—
12	35.0	4 350	870	53	7 000	1 400	50.0	6 940	1 380	48.4	6 420	1 280	—	—	—
14	47.7	5 930	1 180	67	8 800	1 760	68.0	9 440	1 880	65.8	8 730	1 740	—	—	—
16	62.3	7 740	1 540	83	11 000	2 200	88.8	12 300	2 460	86.0	11 400	2 280	92.6	12 300	2 460
18	—	—	—	120	15 800	3 160	—	—	—	109	14 400	2 880	117	15 500	3 100
20	—	—	—	140	18 400	3 680	—	—	—	134	17 800	3 560	145	19 200	3 840
22	—	—	—	163	21 500	4 300	—	—	—	163	21 600	4 320	175	23 200	4 640
24	—	—	—	213	28 000	5 600	—	—	—	193	25 700	5 140	208	27 600	5 520
26	—	—	—	241	31 500	6 300	—	—	—	—	—	—	245	32 400	6 480

Remember: W = weight of cable; BL = breaking load; SWL = safe working load with safety factor 5
* Six-strand cables made of steel wires with a breaking load of 140 kg/mm^2

TABLE 25

Characteristics of two steel chains

Link thickness (mm)	Forged steel chain (steel BL = 37-50 kg/mm²)			High resistance steel chain (steel BL = 90-110 kg/m²)		
	Weight (kg/100 m)	BL (kg)	SWL (kg)	Weight (kg/100 m)	BL (kg)	SWL (kg)
7	—	—	—	108	6 158	1 232
8	145	2 512	1 000	—	—	—
10	225	3 225	1 500	220.7	12 570	2 514
12	325	5 650	2 250	—	—	—
13	—	—	—	372	21 240	4 250
14	440	7 675	3 000	—	—	—
16	575	10 500	4 000	564	32 175	6 435
18	730	12 725	5 000	—	—	—
19	—	—	—	714	45 370	9 000
20	900	15 650	6 200	—	—	—
22	1 090	19 000	7 540	—	—	—
25	1 380	24 550	9 700	—	—	—

Remember: BL = breaking load; SWL = safe working load

14.   Chains consist of a series of interlocked links made of steel such as forged steel or high resistance steel. The thicker the diameter of the steel in the links, the greater the strength of the chain. To a lesser extent, the strength is also affected by the size and shape of the links: short links are stronger but more inclined to kink.

15.   The characteristics of two kinds of chain are given in **Table 25**. When buying a chain, you should always ask about its **steel grade** to be able to compare it with the characteristics given here as examples.

16.   To determine which steel chain to use, find the link thickness required for your kind of work (using **Table 25**) based on the steel grade.

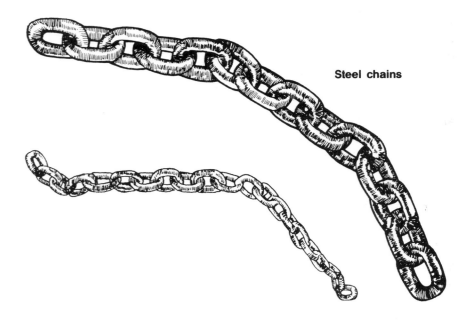

Steel chains

Example

You require SWL = 3 tonnes = 3 000 kg and you plan to buy the cheaper, forged steel chains. From **Table 25**, you can see that you require a link thickness of at least 14 mm. If you use a 10-m chain with 14-mm thick links, it will weigh 440 kg x (10 m ÷ 100 m) = 440 kg x 0.10 = 44 kg.

17.   Chain is preferred to cable in work where the traction line is subject to scraping, kinking or twisting. Chain is not damaged as cable is by sharp bending. It is resistant to abrasion and can be easily attached, detached, lengthened or shortened. It is also easier to repair. Note though that if chain is corroded, badly kinked or worn (e.g. through use as a boat mooring), its strength will be reduced. As a rough guide, you should measure the thinnest part of the chain (after any rust has been knocked off) and estimate its strength accordingly.

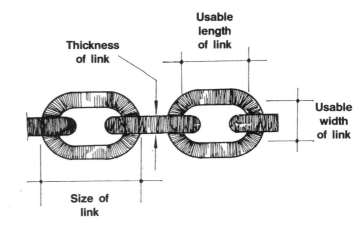

Selecting fittings to use with traction lines

18. In general, traction lines require some sort of fitting on the ends to enable them to be attached to the power source or to the object to be moved.

19. **Cable clamps** are usually standard drop forged clips, which consist of a U-bolt, a saddle and two nuts. The cable is doubled over on itself, and the two thicknesses are squeezed between the U-bolt and the saddle by tightening the two nuts equally. You should use at least two clips per cable and increase their number as the cable size increases.

Note: always use the grooved lower surface of the saddle on the working end of the cable.

20. A metal **thimble or eye** is usually attached at the end of cables to keep them from being damaged. The size of the thimble and the size of its groove should be well matched to the cable size. Cable clamps are used to secure the thimble tightly inside the cable loop.

Drop forged clip

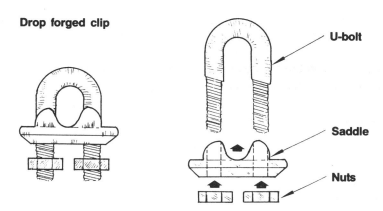

U-bolt

Saddle

Nuts

Double end of cable and attach clips

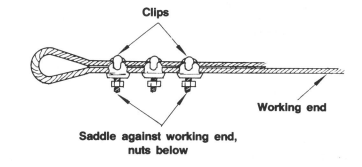

Clips

Working end

Saddle against working end,
nuts below

Metal thimble for loop of cable

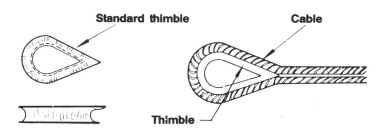

Standard thimble

Cable

Thimble

21. **Hooks** and **rings** are the most popular chain fastenings. Some common types and their uses are listed:

(a) **Slip (round) hook** to form a choker that will tighten under pull.

(b) **Grab hook** to form a chain loop that will not tighten; for example to shorten the chain by lengthening the loop or by blocking the chain so that it will not slide through an opening.

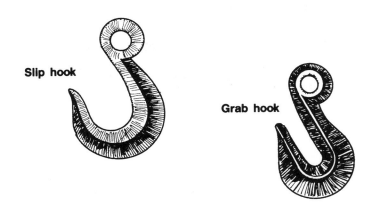

Slip hook

Grab hook

(c) **Rings**, used in the same manner as round hooks, should always be thicker than the chain. They are stronger than hooks. When using two rings on the same chain, one should be narrow enough to slide through the other.

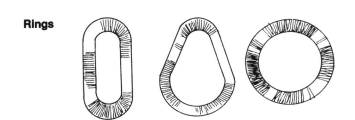

Rings

Using chains with hooks and rings

Logging chain

Grab hook

Slip hook

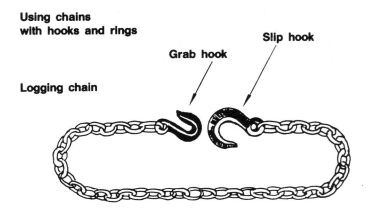

Shortening a chain with a grab hook

Grab hook

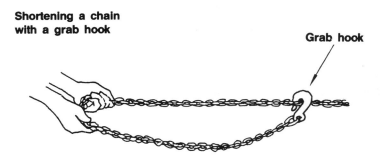

Large ring

Small ring

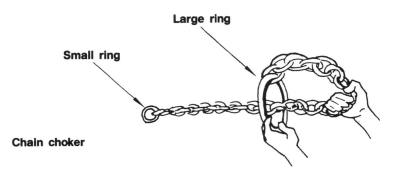

Chain choker

(d) **C-links** are special types of ring, each having a small notched gap through which another C-link can be attached. They are designed for ease of clipping together but can only be separated by the user aligning the gaps correctly. They will not unclip by accident.

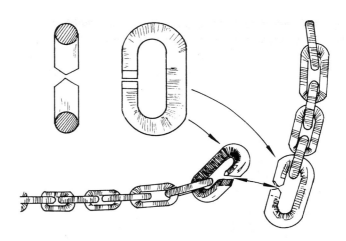

(e) **Quick links** are special rings with screw cylinders which can be opened and closed to connect lengths of chain.

Using a quick link

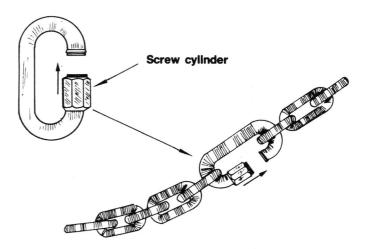

Screw cylinder

22. **Shackles** can be used in place of rings or hooks for many purposes. They are very handy for attaching chains or cables to each other. Be sure to attach them correctly (as shown). They can also be used for emergency chain repair. Characteristics of some standard shackles in forged steel (BL 50 to 65 kg/mm^2) are given in **Table 26** for your guidance.

TABLE 26

**Characteristics of some standard shackles
(forged steel with breaking load 50 to 65 kg/mm^2)**

Shackle thickness (mm)	Opening width (mm)	Breaking load (BL) (kg)	Safe working load (SWL)* (kg)
14	28	7 250	800
16	32	11 000	1 000
18	36	13 200	1 250
20	40	16 000	1 600
22	44	18 000	2 000
24	48	22 000	2 500
27	54	25 000	3 150
30	60	35 000	4 000
33	66	45 000	5 000

* Safety factor greater than 6

Note: when using a screw shackle, close it with the correct pin and screw the pin completely. Then unscrew the pin by half a turn to prevent it from jamming. Make sure the shackle screw is well greased.

Standard shackles

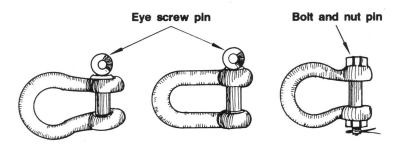

Eye screw pin   Bolt and nut pin

Using a shackle

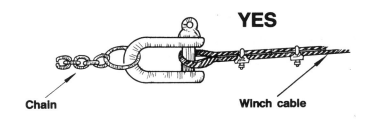

YES

Chain   Winch cable

NO

Chain   Winch cable

210

Single

Double

Triple

23. **A pulley block** is an assembly of one or more pulley wheels, set on an axle that is held in a steel or wooden case. The block can be attached to one or two traction lines, preferably with a swivel hook. **Latched blocks** are similar but are much easier to use, as the cable can be inserted or removed from the side of the block.

Latched

24.   It is important to select the groove of a pulley to match the particular size of the cable to be used. Remember that the groove width should always be a little larger than the cable diameter, never narrower.

The pulley groove diameter should be slightly larger than the cable diameter

Example

Cable diameter (mm)	Pulley groove width (mm)
10	11.0
20	21.5
30	32.0
40	42.5

25.   The use of pulleys is described later (see Section 54, paragraph 10 and following). Pulley blocks should be kept well greased, and pulleys should run free. Flattened or twisted pulleys should be replaced.

Remember: when you are putting together an assembly (such as rope, chain, cable, fittings) the pulling strength will be limited to the weakest component of the assembly.

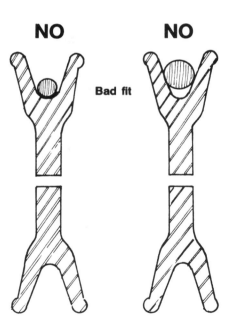

YES

Good fit

NO   **NO**

Bad fit

52 Clearing the site

1. The site should be cleared of all obstructions such as:

   ● woody vegetation, where the roots can cause severe cracking in pond structures such as concrete water inlets and outlets;

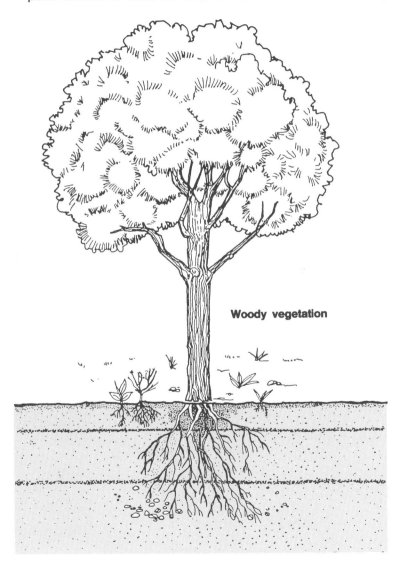

Woody vegetation

   ● tree stumps which, when decaying, can also weaken concrete pond structures by leaving voids in the soil;

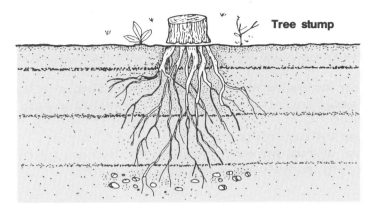

Tree stump

   ● large stones, which may need to be dug out;

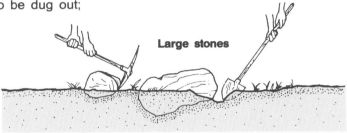

Large stones

   ● ant hills and animal burrows, which should be dug completely; clayey soil should be tramped into the hole created.

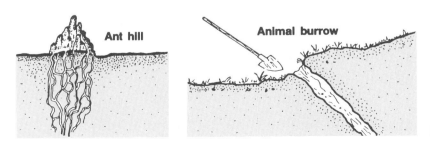

Ant hill   Animal burrow

2.   Define carefully the exact area to be cleared before you begin. Determine the outer corners of the pond area, which should include the entire area to be covered by the dikes (see Section 60). You could mark the area using wooden stakes and cord or poles. When this is done, mark out an additional area beyond the dikes to serve as a work space and a walkway around the site. Then you are ready to proceed.

(a)   **Clear the area within the limit of the pond dikes** of all vegetation, shrubs, trees (including woody roots and tree stumps) and all large stones.

(b)   **Clear the work space** and walkway around the dikes.

(c)   **Clear all trees and shrubs** within 10 m of dikes and pond structures and any access, water supply or drainage area.

Preparing for a pond

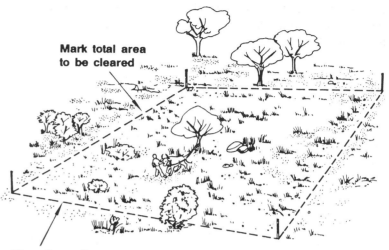

Mark total area
to be cleared

Clear the entire area
including 2 to 3 metres
for work space and walkways
beyond the limit of pond dikes

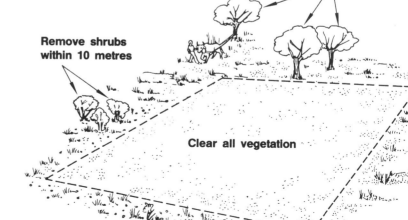

Remove trees
within 10 metres

Remove shrubs
within 10 metres

Clear all vegetation

Note: for the construction of a **barrage pond**, clear all vegetation above ground level on the site of the pond. Then, clear all vegetation including roots from the intended dike area and for a radius of 10 m around the intended drain structure.

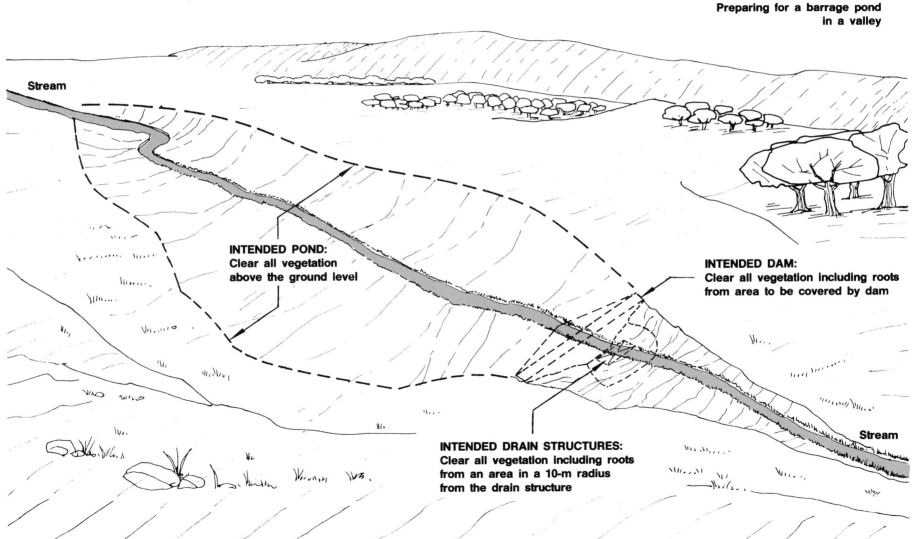

Preparing for a barrage pond
in a valley

Stream

INTENDED POND:
Clear all vegetation
above the ground level

INTENDED DAM:
Clear all vegetation including roots
from area to be covered by dam

INTENDED DRAIN STRUCTURES:
Clear all vegetation including roots
from an area in a 10-m radius
from the drain structure

Stream

3. The clearing method to be used largely depends on the **type of vegetation** on the site. In open savannah country, it is a relatively easy task that can be done manually with limited special equipment. In heavier forested areas on the contrary, clearing generally requires either a large work force and/or the use of machines. If an area is too heavily forested, it might be best not to select it as a construction site, unless no other alternative exists.

4. There are two basic ways to clear the woody vegetation from a site:

● you can cut the trees and then remove the stumps;
● you can fell whole trees with their roots attached.

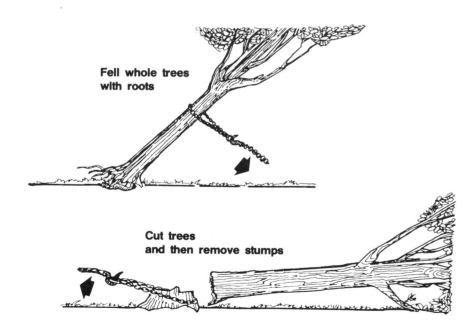

Fell whole trees with roots

Cut trees and then remove stumps

5. You will learn more about these methods in Sections 53 to 55.

6. The clearing of the site should be completed by gathering all cut vegetation, stumps, roots and large stones. All these should be removed from the work area. When conditions are dry enough, the clearing can be completed by starting a fire, which should be carefully kept under control.

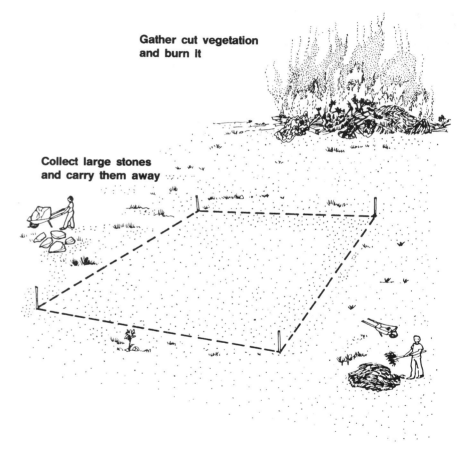

Gather cut vegetation and burn it

Collect large stones and carry them away

Note: you can often sell the wood cleared from the site or make charcoal and sell it at a good price.

53 The cutting of trees

1. Tree cutting is a very dangerous job. It requires much skill and experience to avoid accidents. If you have never cut trees before, you should subcontract this specialized work to people known for their practical experience. You should however learn about a few basic points that will help you to plan this clearing operation better.

2. High stumps are more easily removed than low ones (see Section 54). This is particularly true if you remove the stumps using either a winch or low-powered machinery. Remember that if the wood is valuable, you will get less money from shorter trees.

High stumps are easier to remove, but give less usable timber

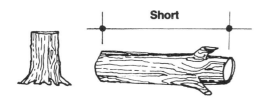

Short

Low stumps are harder to remove, but give more usable timber

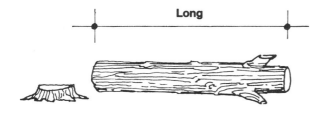

Long

Preliminary work

3. Before starting to cut the trees, you should first **clear out the undergrowth**. This will reduce danger to people by cutting down on tangles with fallen trees.

Remove shrubs and undergrowth

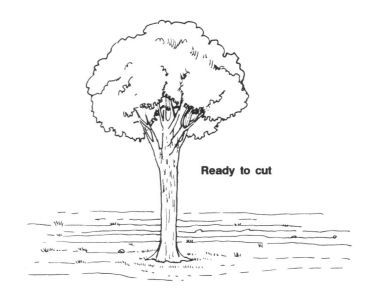

Ready to cut

217

4.   There are two steps when cutting trees.

(a)   **Making the undercut:** cut a pie-shaped piece out of the tree on the side facing the direction in which the tree is supposed to fall. The depth of the cut should be one-fifth to one-quarter of the diameter of the tree.

(b)   **Making the backcut:** 3 to 5 cm above the base of the undercut and at right angles to the falling direction, cut horizontally nearly all the wood that remains on the back of the tree.

5.   If at the end of the backcut, the tree does not fall by itself, hammer a wedge into the backcut to push the tree over in the desired direction. If pulling or guide cables are to be used, they should be attached before the cutting is started.

Note: felling large trees may be dangerous. For trees with a diameter over 60 cm at the stump, special techniques have to be used.

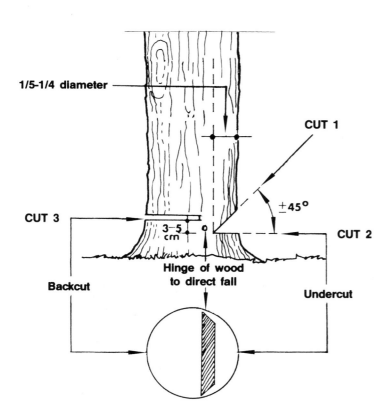

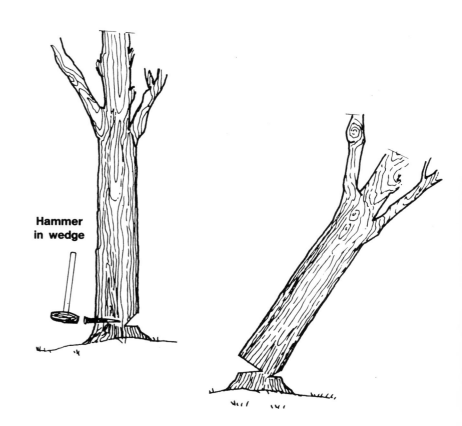

Selecting equipment

6. Small trees can be chopped with an **axe** or they can be sawed. A **handsaw** requires two operators and the use of a wedge in the backcut.

Power saws operated by one person only are increasingly popular. (If you plan to use one, you should refer to **Chainsaws in tropical forests**, *FAO Training Series*, **2**.)

Felling by hand

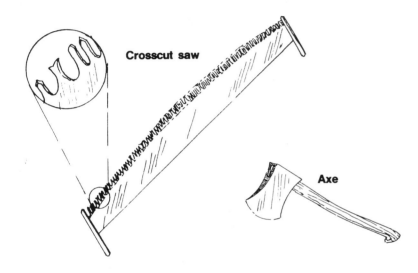

Crosscut saw

Axe

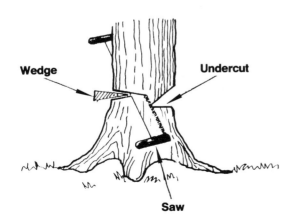

Wedge

Undercut

Saw

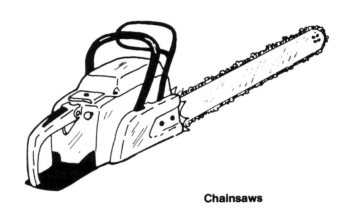

Chainsaws

54 The removal of tree stumps

1. Standard ways of removing tree stumps from the ground are either to dig them out or to pull them out. Small stumps can easily be dug out by hand. Larger stumps may require the use of machine power. Stump pulling is done by placing a rope, cable or chain around the trunk and pulling as follows:

- **direct pull** by an animal or machine;
- **winding in** of a cable on a winch;
- **a combination** of these methods with pulley blocks.

2. In the next sections, you will learn about the most appropriate methods.

Dig small stumps by hand

Pull large stumps with animals, machine power or winch using pulley blocks as needed

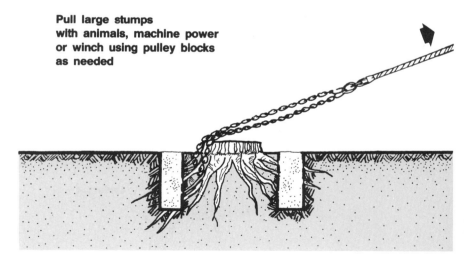

Digging out tree stumps by hand

3. To make work easier, dig trenches around the stump in the following sequence, cutting any roots encountered.

(a) Dig trench 1.

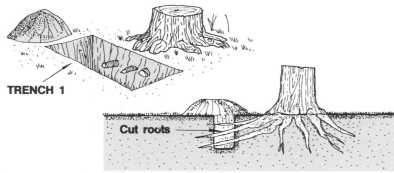

TRENCH 1

Cut roots

(b) Dig trench 2.

TRENCH 2

(c) Dig trench 3.

TRENCH 3

4. Pull the top of the stump as shown. If it fails to be dislodged, dig trench 4 and cut remaining roots. Try again to dislodge the stump.

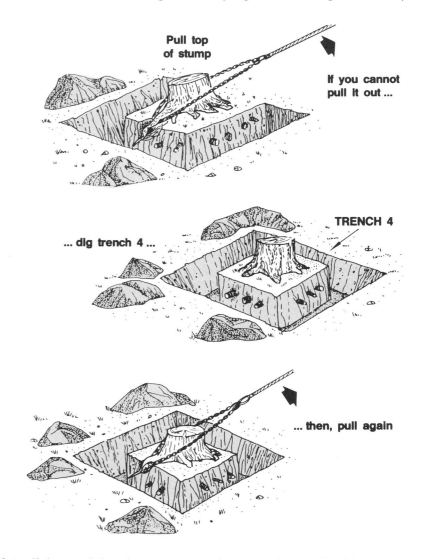

Pull top of stump

If you cannot pull it out ...

... dig trench 4 ...

TRENCH 4

... then, pull again

Note: if the trunk has been cut near the ground, you should excavate all around the stump.

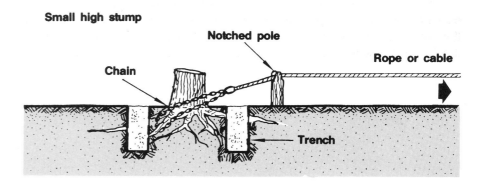

Small high stump

Notched pole

Rope or cable

Chain

Trench

5. Lift the stump out of the ground. If it is a small stump, put a rope or chain around the foot of the stump. If it is a large stump, first attach a rope or chain around one of the strong stump roots.

6. Lead the line over a **notched pole**, a **stumping trestle** or **shearlegs**, which you can easily make yourself. While pulling the line with one group of people, have other people push the stump up with crow bars to free it from the ground.

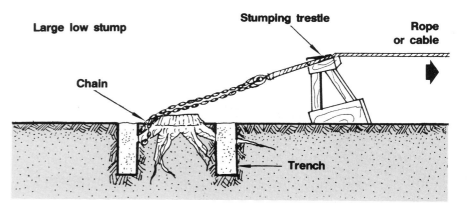

Large low stump

Stumping trestle

Rope or cable

Chain

Trench

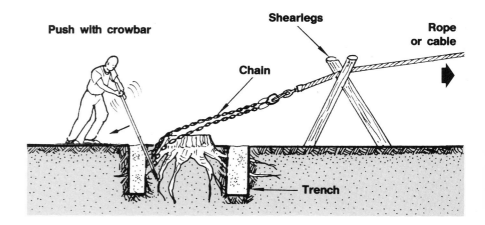

Push with crowbar

Shearlegs

Rope or cable

Chain

Trench

7.   A hand winch consists of **a metal casing** in which **a double set of claws grips a steel cable**. A hook is secured to the back of the casing. As the handle of the winch is cranked, the claws move horizontally and pull the cable inside the winch casing. Excess cable comes out through the back. The cable is equipped with a hook at its end. There are various models of hand winches available, their traction force varying from 0.5 to 10 tonnes. A very useful one is the more common three-tonne hand winch.

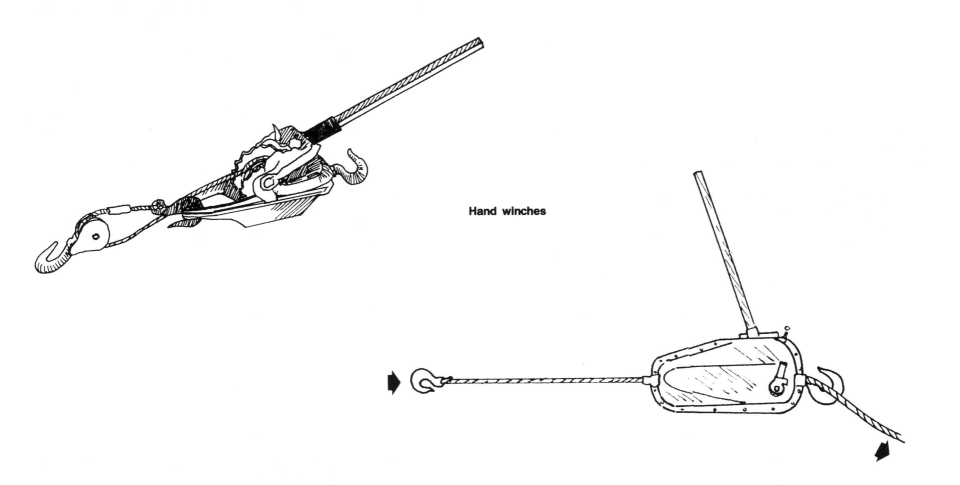

Hand winches

8. To remove a tree stump with a hand winch proceed as follows.

(a) Use a robust standing tree or stump, a rock or a heavy vehicle not too far away from the stump to be removed. The distance between the two should not be longer than the length of the winch cable.

(b) Tie a **sling** (a steel cable with two thimbles) around the base of the standing tree or other object which will be used as an anchor. With a vehicle, attach the hook or cable directly to the towbar or chassis fixing point.

(c) Attach the hand winch to this sling, using the hook at the back of the casing.

Note: if the stump is too far from a possible anchor, enlarge the sling around the anchor with a chain or a cable and connect the latter to the winch hook.

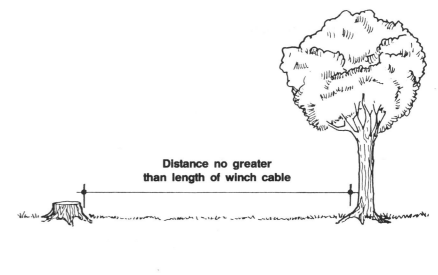

Distance no greater than length of winch cable

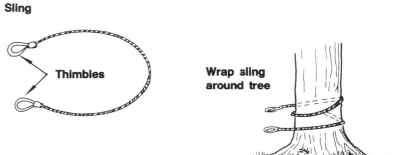

Sling

Thimbles

Wrap sling around tree

Attach winch to sling

Other ways to attach a winch

40 — 50 cm

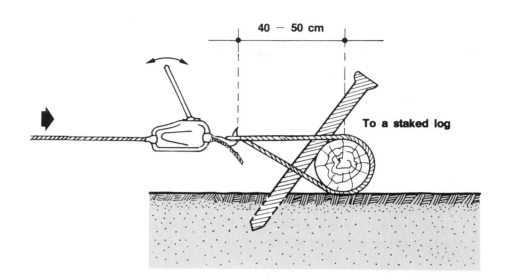

To a staked log

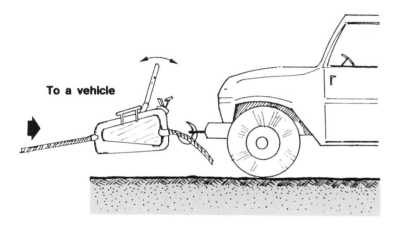

To a vehicle

100 cm

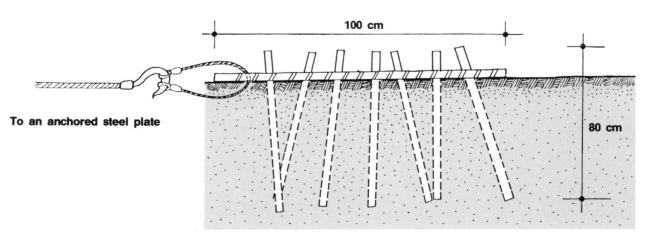

To an anchored steel plate

80 cm

(d) Tie either a sling or a chain around the stump to be removed:

- it is best to use a chain long enough to be wrapped around the stump and form at its end an angle smaller than 60°;
- if the diameter of the stump is too large to do this, it is preferable to use a cable sling;
- to avoid the sling or the chain slipping off the stump, cut a groove around it into which the sling or chain will fit.

Note: using the winch cable as a choker around the stump will damage it and greatly reduce its strength. **Never do it**.

(e) Attach the hook of the winch cable to this sling or chain.
(f) Dig around all or part of the stump so that the bigger roots can be cut.
(g) Slowly start cranking the winch with its handle, checking that all lines and their fittings are in order.

Note: using a winch can be dangerous. Check that each piece of equipment used is strong enough to withstand the traction force applied by the winch. Check regularly for wear and tear. Keep people away from lines under tension.

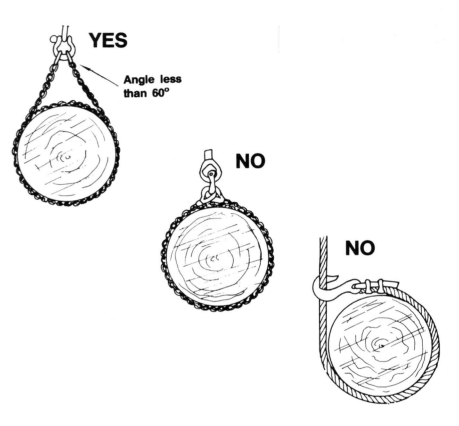

YES

Angle less than 60°

NO

NO

Groove

Using the monkey winch

9.   Another simple, reliable and heavy-duty winch is the **monkey winch**. It can be very effectively used in the same way as the hand winch (see pages 223 to 226).

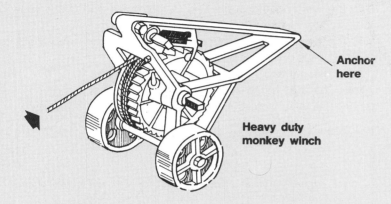

Heavy duty monkey winch

Anchor here

Increasing pulling force

10.   You can easily improve your pulling efficiency for stump removal in the following way.

(a)   Apply a **lifting action** to the stump by using a notched pole, shearlegs or a stumping trestle and attaching the line to one of the strong roots (see page 222).

(b)   Apply a **tipping action** to the stump by making a notch in its top surface and attaching the chain as shown.

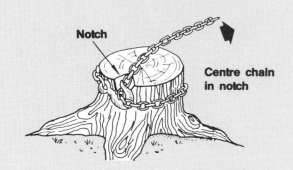

Notch

Centre chain in notch

(c)   **Use pulley blocks** to increase the strength of your winch. Remember that this also increases the loading of the equipment such as slings, shackles and pulley blocks, which should be strong enough to withstand the extra loading (see examples below).

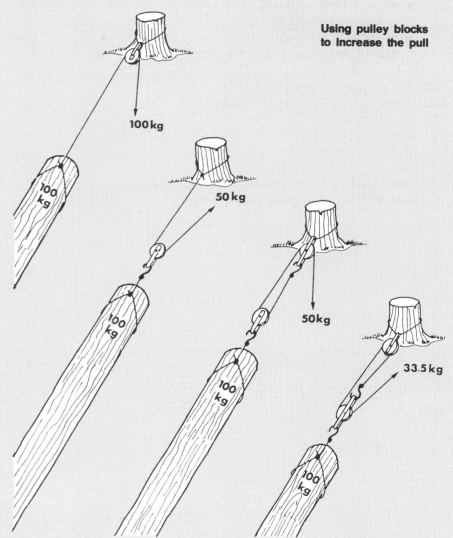

Using pulley blocks to increase the pull

100 kg

100 kg

50 kg

100 kg

50 kg

100 kg

33.5 kg

100 kg

Using the winch with one pulley block

11. To **double the strength of the winch**, you can use one pulley block and one (or better two) anchor(s) as follows:

(a) Attach the winch to its anchor (see paragraph 8).
(b) Run the winch cable in the block over the pulley and run it back to the anchor.
(c) Attach the winch cable to this anchor (see paragraph 8).
(d) Using a cable or a chain, attach the hook of the pulley block to the stump to be removed.
(e) Dig a trench around the stump and cut all major roots that can be reached close to the stump.
(f) Crank the winch carefully.

Example

If you are using a small one-tonne winch, you can now pull the stump out with a two-tonne traction force as follows:

- traction on winch cable = 1 tonne
- traction on single anchor = 1 + 1 = 2 tonnes
- traction on stump = 2 tonnes

Note: if a **second anchor** is available, you may use it to attach the end of the winch cable; this will reduce the traction force by half on each of the two anchors, as shown on the opposite page.

Using the winch with two pulley blocks

12. **To treble the strength of the winch**, you can use two pulley blocks and two anchors, as follows:

(a) Attach the winch to anchor 1.
(b) Run the winch cable in the first pulley block and run it back toward the second block.
(c) Run the winch cable in this second pulley block, run it back toward the first block, and attach the cable to it.
(d) Attach the second block to strong anchor 2.
(e) Attach the first block to the stump.
(f) Crank the winch carefully, checking anchor 2 especially.

Example

If you are using a small one-tonne winch, you can now pull the stump out with a three-tonne traction force as follows:

- traction on winch cable = 1 tonne
- traction on anchor 1 = 1 tonne
- traction on anchor 2 = 2 tonnes
- traction on stump = 3 tonnes

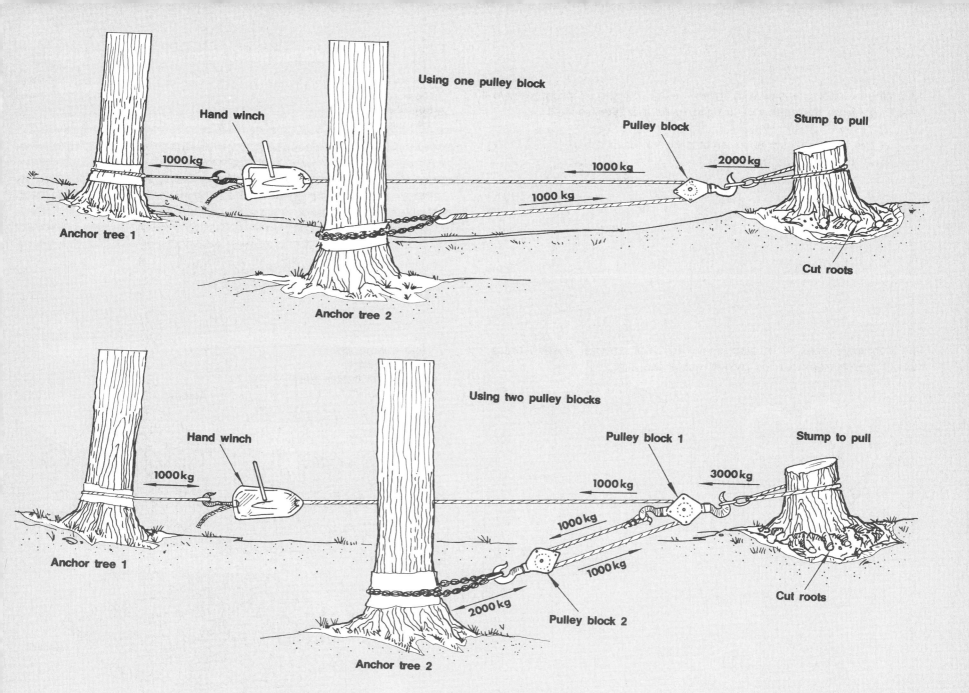

Using one pulley block

Hand winch

Pulley block

Stump to pull

1000 kg

1000 kg

2000 kg

1000 kg

Anchor tree 1

Anchor tree 2

Cut roots

Using two pulley blocks

Hand winch

Pulley block 1

Stump to pull

1000 kg

1000 kg

3000 kg

1000 kg

1000 kg

2000 kg

Pulley block 2

Anchor tree 1

Anchor tree 2

Cut roots

13. When clearing a site with many trees, you should progress systematically **from the centre of the site to its margins**, so that:

- the trees can serve as an anchor for each other;
- all trees can be pulled out.

Clear a small site from the centre with one access way

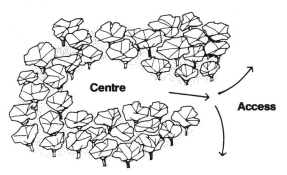

14. For large sites, it is also useful to clear several **access areas** through which trees can be moved out of the site.

Clear a large site from the centre with several access ways

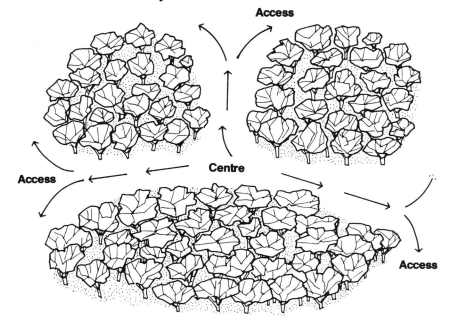

Removing stumps with a bulldozer

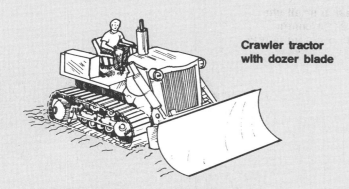

Crawler tractor with dozer blade

15. If a tractor equipped with a front-mounted **dozer blade** is available, it can considerably speed up the removal of stumps. This is particularly important for a forested site. For best efficiency, proceed as follows:

(a) **Raise the dozer blade** and place it against the trunk part of the stump.

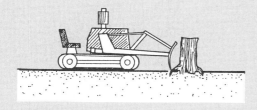

(b) As you push against the stump, raise the blade further **to apply a lifting action** to the stump; it will also increase the grip of the tractor on the ground.

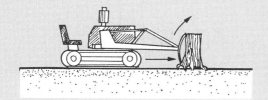

(c) **Tilt the trunk forward** until the roots in front of the tractor are slightly out of the ground.

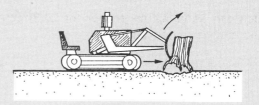

(d) Back the tractor and **lower the blade**.

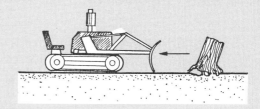

(e) **Insert the blade** under the roots.

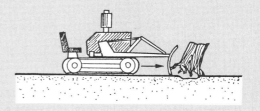

(f) **Push** while lifting with the blade to break the stump out of the ground.

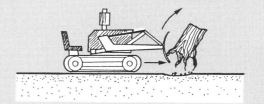

55 Uprooting whole trees

1.   A common method of removing smaller trees and their roots from a construction site is to uproot whole trees. This can be relatively easy because of the greater leverage obtained when pushing or pulling a tree at a point high up its trunk. But to be safe, **this method should be used only if:**

- there is adequate space;
- the tree is not rotten, cracked or split.

Note: the pulling angle with the horizontal should not exceed 30°; if space is restricted or a greater angle of pull is required, **use a pulley block to change the pulling direction.**

Using a hand winch to uproot a whole tree

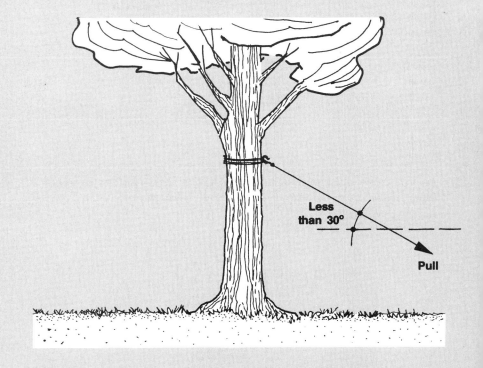

Uprooting a whole tree using a winch

2. If you only have a hand winch or a monkey winch to work with, you can proceed as follows:

(a) Position the winch at **a distance at least twice the height of the tree** and as close as possible to a robust tree that can be used as the anchor.

(b) Attach the back of the winch to a sling/chain tied around the base of this anchor tree.

(c) Climb up the tree to be uprooted and attach **the winch cable high up the tree trunk** to a sling/chain tied around it.

(d) **Dig a trench** around the base of the tree to reduce the resistance of the stump against pulling.

(e) **Place a log** against the base of the tree facing the winch to assist in the extraction of a maximum amount of roots and aid in removal of the soil.

(f) **Crank the winch** carefully.

Note: you can also use two anchor trees, for example to change the direction in which the uprooted trees will fall.

Using one anchor tree

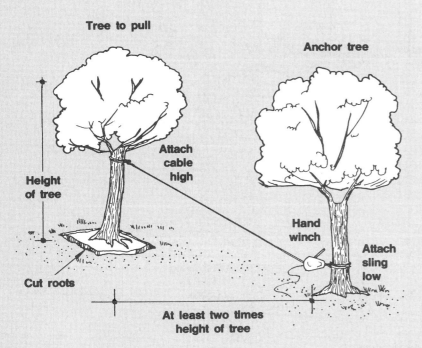

Tree to pull

Anchor tree

Attach cable high

Height of tree

Hand winch

Attach sling low

Cut roots

At least two times height of tree

Using two anchor trees

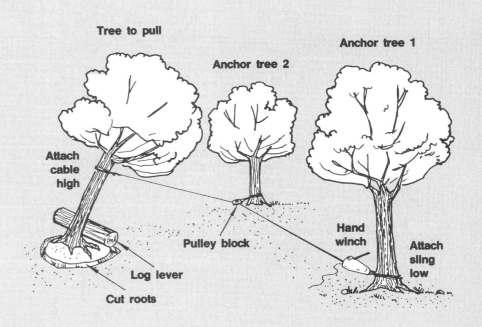

Tree to pull

Anchor tree 2

Anchor tree 1

Attach cable high

Pulley block

Hand winch

Attach sling low

Log lever

Cut roots

Note: always attach the cable to the tree to be pulled as high as possible and the sling to the anchor tree as low as possible

233

Uprooting a whole tree using a winch and pulley blocks

3. It will be much easier to uproot a tree if you **increase the strength of the winch** by using **two pulley blocks** for example. This will treble the pulling force of your winch (see Section 54).

4. Remember that in this particular case, you will need a particularly **strong anchor tree**.

Using two pulley blocks

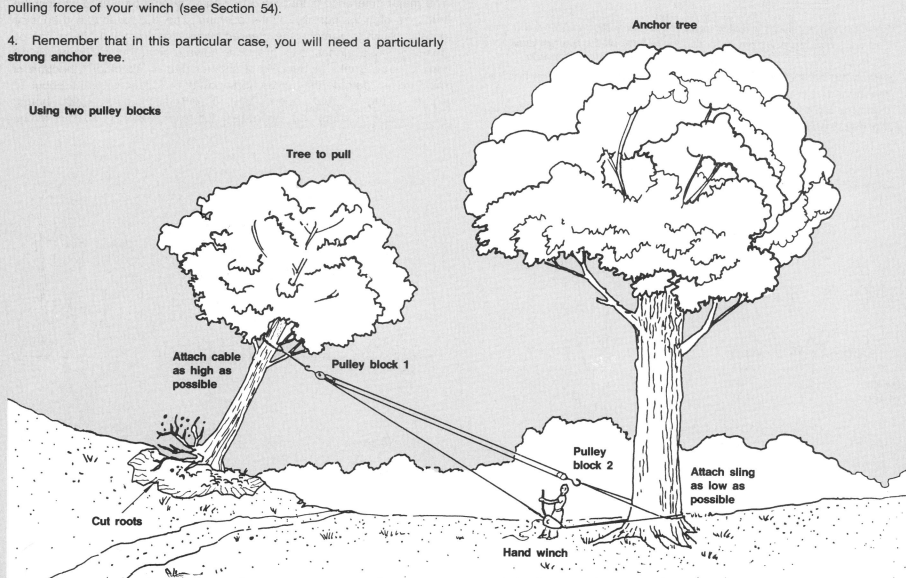

Anchor tree

Tree to pull

Attach cable as high as possible

Pulley block 1

Cut roots

Pulley block 2

Attach sling as low as possible

Hand winch

Using a bulldozer

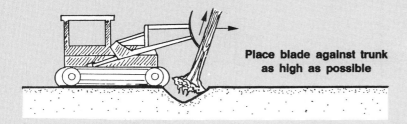

Place blade against trunk as high as possible

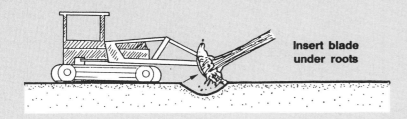

Insert blade under roots

Push out branches and rubbish

Towing out large trunks

5.   If **a bulldozer** is available, it can be used to uproot trees, in a way similar to the one described for the removal of stumps (see Section 54). The major difference is that **the dozer blade should be placed against the trunk as high as possible** while pushing. The bulldozer can then also greatly assist to remove the cleared vegetation from the site either by pushing loose stones and branches toward a collection point, or by towing large trunks or bundles of smaller pieces. To clear 1 hectare of medium to heavy forest, an 80 horsepower bulldozer will take about 10 hours. Using manual labour only, it will take about 1 000 hours. Additional work outputs of bulldozers for bush clearing and tree felling are given in **Table 19**.

Track-loader with backhoe and bucket

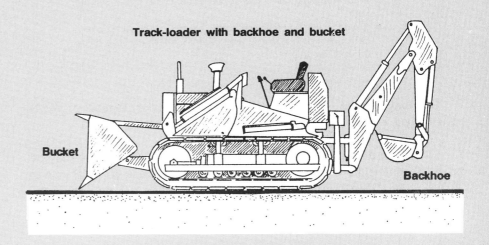

Bucket

Backhoe

6.   **A track-loader** can also be very effective for uprooting live trees, mostly because of its combination of tractive force, high reach of the bucket arms and hydraulic force. It should be used in the same way as a bulldozer. **An excavator** can also be used for smaller trees and can be useful for cutting around roots. You can also use it for lifting bundles of cleared wood.

56   Surface soil removal

1.   Surface soil has the highest concentration of roots and decaying organic materials (see Section 1, **Soil, 6**). This soil is unstable as a construction material and cannot be used for the foundations of any dike or structure. Therefore the surface soil should be removed from the areas where:

- dikes and structures will be built;
- soil will be taken as a dike construction material.

Example:

- If you build a **barrage pond**, remove the surface soil from the area where the dike (and the outlet structure, if any) will be built.
- If you build a **diversion pond** by mixed excavation/embankment, remove the surface soil from the whole area of the pond, dikes and structures included.

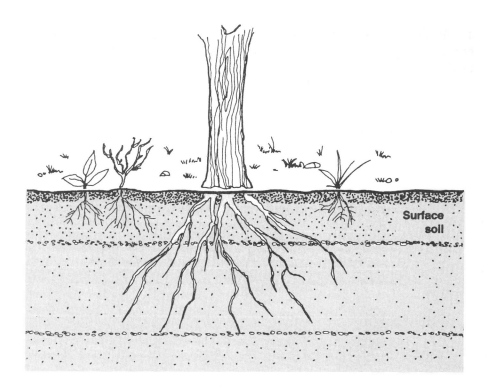

Surface soil

2.   Soil may also be required outside the site to **supply topsoil** for newly constructed areas such as dikes.

3.   **The depth of the surface soil** varies from region to region. It may be totally absent or more than 1 m thick. Usually the surface soil is from **5 to 30 cm deep**. Once your site has been cleared, find out how thick the surface soil is (see **Soil, 6**). On this basis plan the construction method for your dikes and the removal and storage of the surface soil.

Surface soil is usually from 5 to 30 cm deep

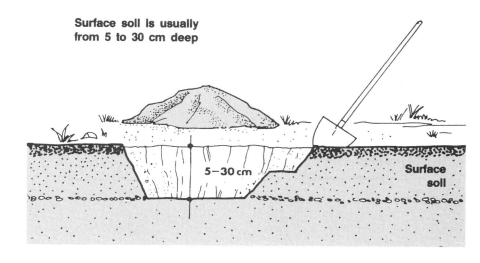

5–30 cm

Surface soil

Example:

- The surface soil thickness averages a few centimetres only: you do not have to remove it, but you will have to plough the area well where the dikes will be built.
- The thickness of the surface soil averages 20 cm over a site of 20 metres square (400 m²): you will have to remove, transport and store 20 m × 20 m × 0.20 m = 80 m³ of soil.

4. **Stake out** clearly the areas of the site from where the surface soil should be removed, as was done earlier before the clearing started (see Section 52).

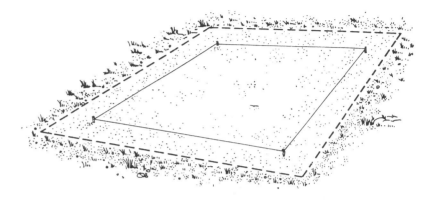

5. **Excavate** to the desired depth and **transport** the surface soil away from the construction site (see Chapter 4).

Remove the surface soil

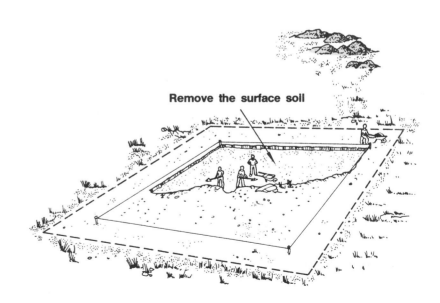

Loosen the soil

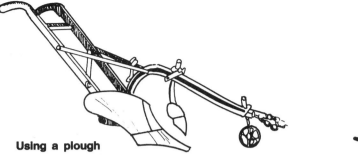

Using a plough

Using a pick

6. To be able to dig out this surface soil without too much effort, it might be necessary to **loosen it by ploughing**. In exceptional cases, you may have **to rip the soil** with a tractor first, before starting excavation. Use as many shanks as possible, at low speed and maximum soil penetration, to reach best efficiency. For small areas, the soil may be loosened using a pick.

Shank

Using a tractor

7. **Store this surface soil** in a suitable location, as close as possible to the site. You will **use this fertile organic soil later** for several purposes such as:

- covering the top and dry side of the dikes with a thin layer of rich soil to grow a protective grass cover (see Section 69);

Cover top and dry sides of dike

- putting it back into the pond to increase its fertility;

Cover pond bottom

- preparing compost piles;

Compost pile

- improving your garden and producing valuable crops.

Your garden

Note: larger sites may present considerable variation in surface soil depth. It is useful to measure this and plan the excavation and movement of the soil accordingly.

239

6 FISH POND CONSTRUCTION

60 Introduction

1.   When the construction site has been prepared, the fish pond and its water control structures can be built. This chapter shows you how to construct the fish pond, while the next book of this manual, **Pond construction**, **20/2**, deals with water control structures.

2.   Dikes are the most important part of a fish pond, as they keep the necessary volume of water impounded and form the actual pond; their design and construction is particularly important. You will learn more about pond dikes and earthwork calculations in the next three sections, before learning how to stake out and construct the four main types of pond.

3.   You will find it useful to have a notebook in which to make any calculations required and, if available, some squared or graph paper for sketching out and measuring pond and dike shapes.

61 Characteristics of pond dikes

1.   Any pond dike should have three basic qualities.

(a)   It should be able to **resist the water pressure** resulting from the pond water depth.

(b)   It should be **impervious**, the water seepage through the dike being kept to a minimum.

(c)   It should be **high enough** to keep the pond water from ever running over its top, which would rapidly destroy the dike.

Resisting water pressure

2.   Water pressure can be readily resisted by:

● **anchoring** your dike strongly to its foundations (the soil on which you build it);

● constructing your dike large enough to **resist the water pressure** by virtue of its weight.

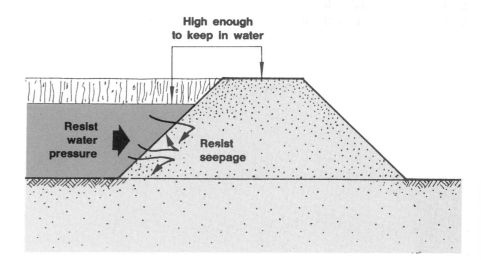

Some points to remember about dikes for good pond dike construction

High enough to keep in water

Resist water pressure

Resist seepage

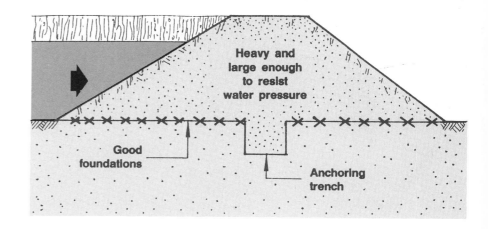

Heavy and large enough to resist water pressure

Good foundations

Anchoring trench

Note: an intermediate dike separating two ponds may not need to be as strong as a perimeter dike, so long as the water pressure is more or less the same on both sides. If one pond needs to be drained while the adjacent one remains filled however, water pressures will be similar to those in perimeter dikes, and the dike should be stronger.

Sections of dikes between ponds and perimeter dikes

Dikes between ponds

Perimeter dikes

Equal water pressure

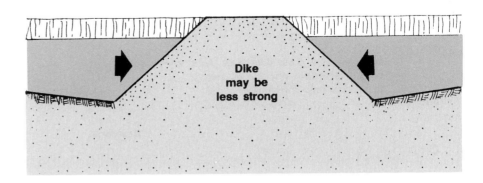

Dike may be less strong

Unequal water pressure

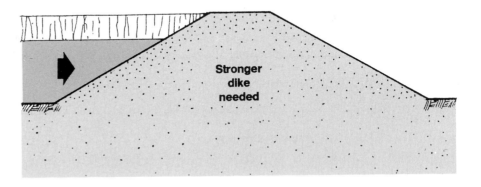

Stronger dike needed

243

Ensuring impermeability

3. Impermeability of the dike can be ensured by:

- using good soil that contains enough clay (see **Soil**, **6**);
- building a central **clayey core** when using pervious soil material;
- building a **cut-off trench** when the foundation is permeable;
- applying good construction practices (see Section 62);
- ensuring that the thickness of your dike is appropriate.

Diagram of a pond dike built using sandy soil with a clay core and cut-off trench to ensure impermeability

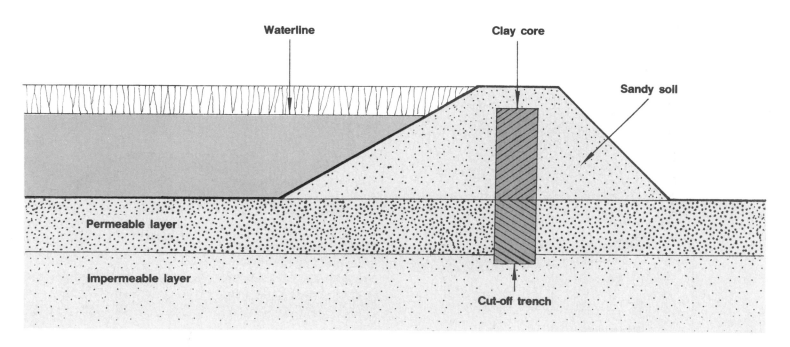

Good dike

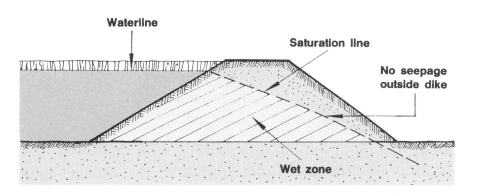

Waterline

Saturation line

No seepage outside dike

Wet zone

The better the soil the thinner the dike

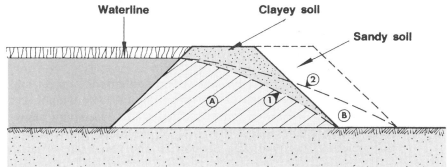

Waterline

Clayey soil

Sandy soil

A   Wet zone of dike in clayey soil
1   Saturation line
B   Wet zone of dike in sandy soil
2   Saturation line

Note: a dike entirely built of good soil is said to be impervious when the upper limit of its wetted zone, the **saturation line*** progresses through the dike so as to remain inside. The better the soil for dike construction, the more the saturation line is deflected downward, and the thinner the dike can be. The slope of this saturation line, the **hydraulic gradient***, usually varies from 4:1 (clayey soil) to 8:1 (sandy soil). As shown, a clay core will affect this hydraulic gradient.

Bad dike

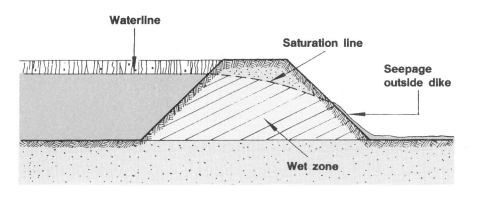

Waterline

Saturation line

Seepage outside dike

Wet zone

Clay core lowers saturation line

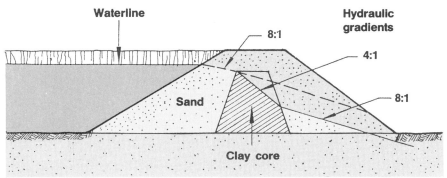

Waterline

Hydraulic gradients

8:1

4:1

8:1

Sand

Clay core

245

4. To calculate the height of the dike to be built, take into account:

- the depth of the water you want in the pond;
- the **freeboard***, which is the upper part of a dike and should never be under water. It varies from 0.25 m for very small diversion ponds to 1 m for barrage ponds without a diversion canal;
- the dike height that will be lost during **settlement***, taking into account the compression of the subsoil by the dike weight and the settling of fresh soil material. This is the **settlement allowance** which usually varies from 5 to 20 percent of the construction height of the dike (see Section 62 and **Table 28**).

Factors to be considered in calculating dike heights

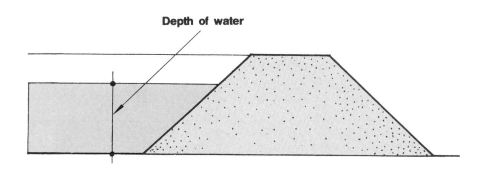

Depth of water

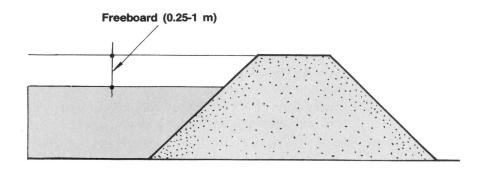

Freeboard (0.25-1 m)

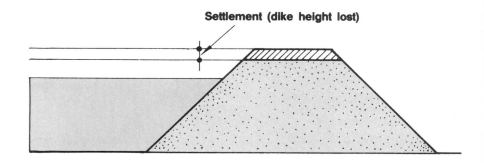

Settlement (dike height lost)

5. Accordingly, two types of dike height may be defined:

- the **design height DH**, which is the height the dike should have after settling down to safely provide the necessary water depth in the pond. It is obtained by adding the water depth and the freeboard;
- the **construction height CH**, which is the height the dike should have when newly built and before any settlement takes place. It is equal to the design height plus the settlement height.

6. You can determine the construction height (**CH** in m) simply from the design height (**DH** in m) and the settlement allowance (**SA** in percent) as follows:

$$CH = DH \div [(100 - SA) \div 100]$$

Example

If the maximum water depth in a diversion pond of medium size is 1 m and the freeboard* 0.3 m, the design height of the dike will be DH = 1 m + 0.30 m = 1.30 m. If the settlement allowance is estimated to be 15 percent, the required construction height will be CH = 1.30 m ÷ [(100 − 15) ÷ 100] = 1.30 m ÷ 0.85 = 1.53 m.

Design height and construction height

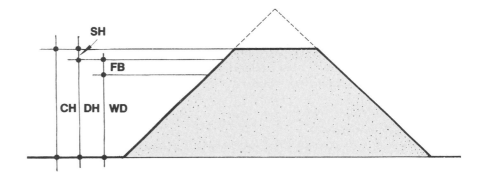

WD = Water depth
FB = Freeboard
DH = Design height
SH = Settlement height
CH = Construction height

Calculating construction height (diversion pond)

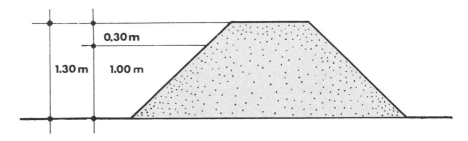

1.30 m ÷ [(100 − 15) ÷ 100] = 1.30 m ÷ 0.85 = 1.53 m

7. **In barrage ponds with a spillway**, the design height of the dike is calculated slightly differently (see Sections 113 and 114, **Pond construction**, 20/2), the freeboard being added above the **maximum level of the water** in the discharging spillway.

Note: the water surface in your pond will be horizontal and therefore the top of your dike should also be horizontal, from the deepest point of the pond to its shallowest point.

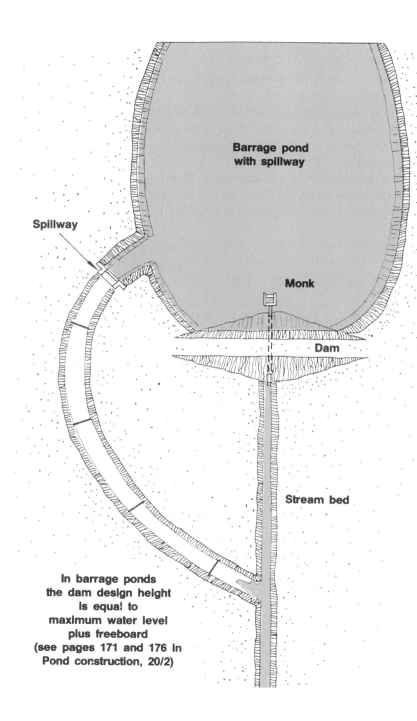

Barrage pond with spillway

Spillway

Monk

Dam

Stream bed

In barrage ponds the dam design height is equal to maximum water level plus freeboard (see pages 171 and 176 in Pond construction, 20/2)

Determining dike thickness

8. A dike rests on its base. It should taper upward to the dike top, also called the **crest** or crown. The thickness of the dike thus depends on:

- the width of the crest; and
- the slope of its two sides.

9. This, together with the height of the dike, will determine the width of the **dike base** (see **Table 27** for examples).

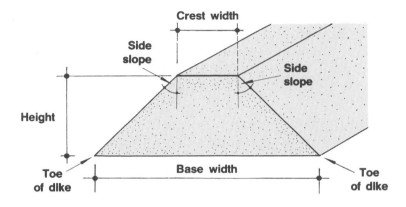

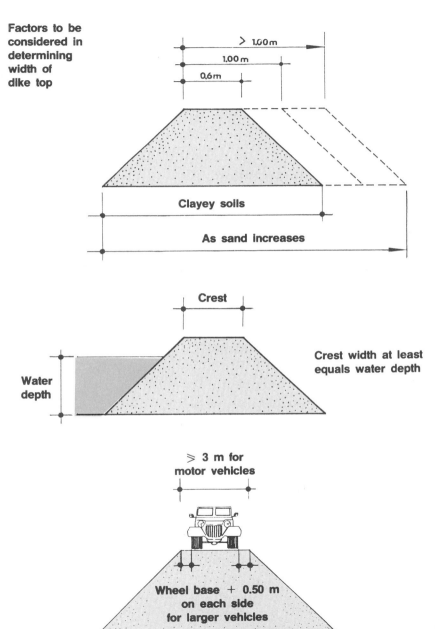

10. Determine the **width of the crest** according to the water depth and the role the dike will play for transit and/or transport.

(a) It should be at least equal to the water depth, but not less than 0.60 m in clayey soil or 1 m in somewhat sandy soil.

(b) It should be even wider as the amount of sand in the soil increases.

(c) It should be safe for the transport you plan to use over it:

- at least 3 m for motor vehicles;
- for larger vehicles at least the wheel base plus 0.50 m on each side.

Note: these dimensions may be slightly reduced for very small rural ponds.

249

TABLE 27

Examples of dimensions of dikes

Individual pond size (m^2)	200		400-600		1 000-2 500	
Quality of soil[1]	Good	Fair	Good	Fair	Good	Fair
Water depth (max. m)	0.80		1.00		1.30	
Freeboard (m)	0.25		0.30		0.50	
Height of dike[2] (m)	1.05		1.30		1.80	
Top width[3] (m)	0.60	0.80	1.00	1.30	1.50	2.00
Dry side, slope (SD)	1.5:1	2:1	1.5:1	2:1	1.5:1	2.5:1
Wet side, slope (SW)	1.5:1	2:1	2:1	2.5:1	2:1	3:1
Base width[4] (m)	4.53	6.04	6.36	8.19	8.92	13.66
Settlement allowance (%)	20	20	15	15	15	15
Construction height[5] (m)	1.31	1.31	1.53	1.53	2.12	2.12
Cross-section area (m^2) / Volume per linear m (m^3)	3.3602	4.4802	5.6266	7.2560	11.0452	16.5996

[1] See **Soil**, **6**. Good soil includes clay, sandy clay, sandy clay loam, clayey loam, silty clay and silty clay loam; fair soil includes loam, sandy loam and silty loam
[2] Design height, the height the dike should have after settlement
[3] To be increased for use of motorized transport
[4] At construction, and calculated from the construction height
[5] Height at which the dike should be built taking settlement into account

11. Not all the dikes of your fish farm are to be used by vehicles (see Section 19). But additional dike width may be required at turning points, based on the diameter of the turning circle of the vehicle used:

- about 3 m for a two-wheel mini-tractor;
- about 4 m for a standard agricultural tractor;
- about 11 m for a light pickup truck;
- additional space would be needed for turning trailers.

Diameters of turning circles needed for the use of various vehicles on the top of dikes

Two-wheel tractor

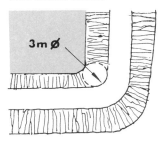

3 m ⌀

Standard tractor

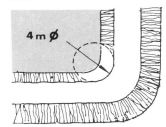

4 m ⌀

Light pickup truck

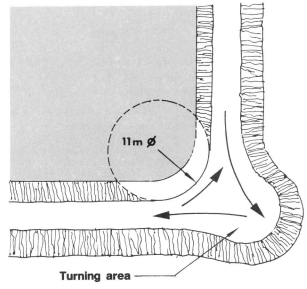

11 m ⌀

Turning area

251

12. In individual ponds, dikes have two faces, **the wet side** inside the pond and **the dry side** or external side. These two sides should taper from the base to the top at an angle that is usually expressed as a ratio defining the change in horizontal distance (in m) per metre of vertical distance as, for example, 2:1 or 1.5:1 (see **Topography**, **16/2**).

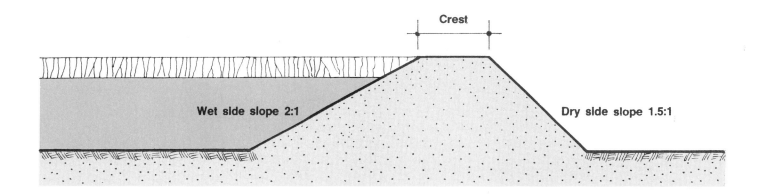

Example

In a dike with side slope 2:1, for each 1 m of height, the base width increases on each side by 2 x 1 m = 2 m.

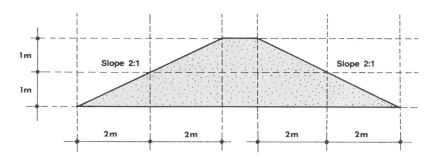

Note: to express the side slope of dikes in other ways, you can use the chart provided here:

Slope (ratio)	Slope (%)	Slope (degrees, approximate)
1:1	100	45
1.5:1	66	34
2:1	50	27
2.5:1	40	22
3:1	33	18

13. The **side slopes** of each dike should be determined bearing in mind that:

- the steeper the slope, the more easily it can be damaged;
- as the soil becomes more sandy, its strength decreases, and slopes should be more gentle;
- as the size of the pond increases, the size of the waves increases and erosion becomes stronger;
- as the slope ratio increases, the volume of earthwork increases, and the overall land area required for the ponds increases;
- a higher slope ratio makes it easier when using a bulldozer to build the dikes.

14. Usually side slopes of dikes vary from 1.5:1 to 3:1 depending on local conditions. The **slope of the dry side can be made steeper** than the slope of the wet side. (see **Table 27** for various sizes of fish ponds and two groups of soil.)

15. In some cases, you may wish to change the slope, for example:

- to provide an easily accessible area for harvesting or operating a monk (see Section 107, **Pond construction**, **20/2**).
- to deepen the pond near the edges to discourage weed growth or bird predation;
- to make edges shallower for good fry feeding.

16. However, you may have to spend more time in maintaining these dikes.

Remember: the thickness of intermediate dikes can be reduced when resistance to water pressure and impermeability are less important.

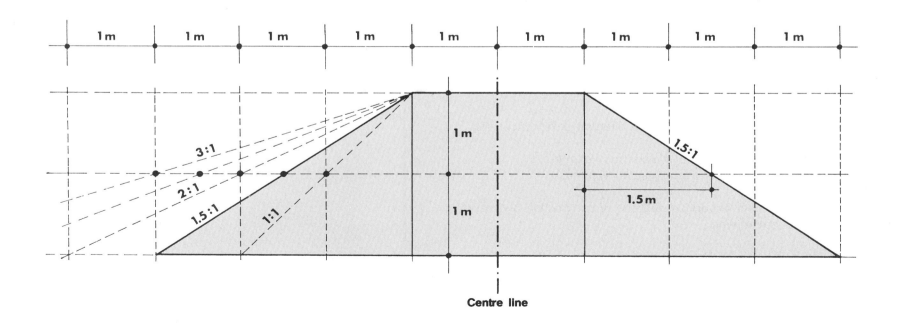

Centre line

62 Compacting earthen dikes

Expansion, compaction and settlement of soils

1. When earth is **disturbed**, for example when it is excavated in preparation for using it to construct dikes, it normally becomes looser, more permeable and less stable. Its volume expands, which is known as **bulking**.

2. When disturbed earth is **compacted**, for example during dike construction, its volume decreases. Later, as the soil settles, the volume is further reduced.

3. Therefore several different but related measurements of earth volumes can be defined:

- the **undisturbed volume**, which is the volume of the soil at the site before excavation;
- the **expanded volume**, which is the volume of the soil after it has been dug out, typically 5 to 25 percent more than the undisturbed volume (see **Table 28**);
- the **construction volume**, which is the volume of the soil necessary for building the dike, for example before any compaction or settlement. It is about the same as the expanded volume;
- the **design volume**, which is the intended volume of the dike after it has been compacted and it has fully settled. Typically, it is 10 to 25 percent less than the construction volume.

4. **Table 28** shows typical characteristics of different soils concerning:

- **expansion**, in percent of the undisturbed volume;
- **settlement**, in percent of the expanded volume.

5. It also shows the effect of various degrees of compaction and rain exposure on further settlement.

6. The primary objectives of **dike compaction** are to start the settling of freshly placed earth, to reduce water permeability, and to strengthen the dike to keep any part of it from sliding away (see also Section 122, **Soil**, **6**).

254

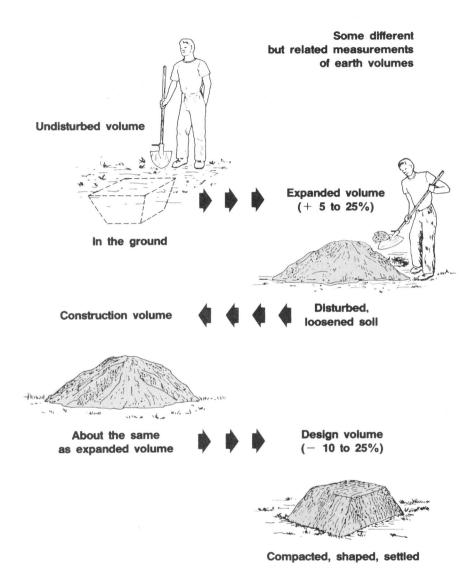

Some different but related measurements of earth volumes

Undisturbed volume

In the ground

Expanded volume (+ 5 to 25%)

Disturbed, loosened soil

Construction volume

About the same as expanded volume

Design volume (− 10 to 25%)

Compacted, shaped, settled

Note: as a rough estimate, considering that expansion and compaction/settlement factors are similar, the undisturbed volume equals the design volume.

TABLE 28

Expansion and settlement of pond soils

Type of soil	Expansion when disturbed (% undisturbed volume)	Settlement allowance* (% expanded volume)
Loose rock, gravel	10-15	8-10
Firm compact earth	10-20	10-15
Ordinary loose earth	5-10	15-20
Loam to light clay	15-25	20-25
Heavy clay	5-15	15-25
Compaction		*Further expected settlement (% construction volume)*
Good: soil layered/rolled/watered		1-5
Normal: soil rolled/watered		5-10
Poor: soil rolled/not watered		10-15
Poorly compacted soils after exposure		*Further expected settlement (% construction volume)*
One rainy season		8-12
Two rainy seasons		5-10
Three rainy seasons		2-5

* Total amount by which the expanded soil volume is expected to be reduced by either compaction plus final/small settlement, or compaction and settlement, or settlement alone

Determining the potential for compaction

7.   You can estimate the bulking of any earth material and determine its potential for compaction by measuring out a known volume of material on the site to be excavated, digging down to the intended excavation depth if possible. You can then either measure the earth volume (for example with buckets, boxes, etc.) or fill the earth back into the test hole and measure the surplus. You should then be able to compact at least 80 percent of this surplus back into the hole by ramming or stamping the soil.

Example

Using a 0.30 × 1 m trench, dig to a 1 m depth. The original earth volume = 0.30 m³. The earth is filled back, leaving a surplus of 0.06 m³ or 60 l.

(a)   Estimate bulking

● expanded volume = 0.30 m³ + 0.06 m³ = 0.36 m³
● bulking (in percent) is obtained as:
[(expanded volume − undisturbed volume)
÷ undisturbed volume] × 100
= [(0.36 m³ − 0.30 m³) ÷ 0.30 m³] × 100
= (0.06 m³ ÷ 0.30 m³) × 100 = 20 percent.

(b)   You should expect to be able to compact at least 80 percent of the surplus (the difference between the expanded volume and the original undisturbed volume): 0.06 m³ × 0.80 = 0.05 m³. The compaction potential is calculated as: (0.05 m³ ÷ expanded volume) × 100 = (0.05 m³ ÷ 0.36 m³) × 100 = 13.9 percent of the expanded volume.

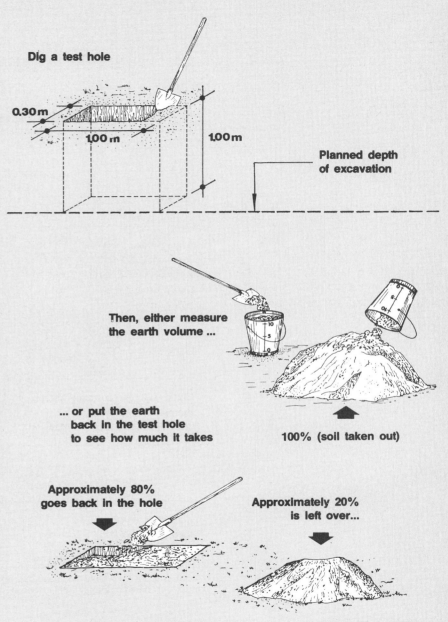

Dig a test hole

0.30 m

1.00 m   1.00 m

Planned depth of excavation

Then, either measure the earth volume ...

... or put the earth back in the test hole to see how much it takes

100% (soil taken out)

Approximately 80% goes back in the hole

Approximately 20% is left over...

... and 80% of this can be compacted back into the hole

8. If the site soil was initially loose, you may be able to compact it into a smaller volume than the original one. To define the **compaction potential**, you can then measure the loose earth required to fill the hole back up to the original level.

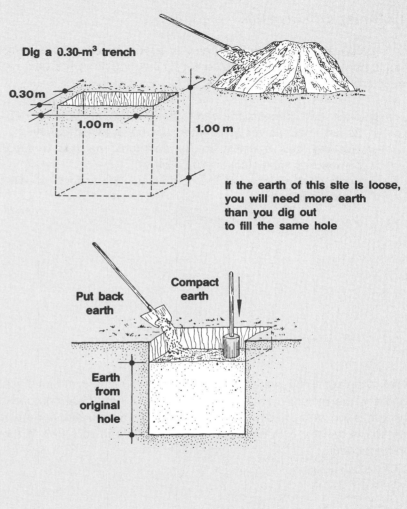

Dig a 0.30-m³ trench

0.30 m

1.00 m

1.00 m

If the earth of this site is loose, you will need more earth than you dig out to fill the same hole

Example

A 0.30-m³ trench is dug and the soil is stamped back, requiring 0.06 m³ of loose fill to make it level. The compaction potential of the original soil equals (0.06 m³ ÷ 0.30 m³) x 100 = 20 percent.

9. Make note of the basis on which the above calculations are made, that is, whether the compaction potential is related to the expanded soil volume or the original soil volume. **Be certain to understand the relationships between undisturbed, expanded, construction and design volumes, as explained on page 254.**

Put back earth

Compact earth

Earth from original hole

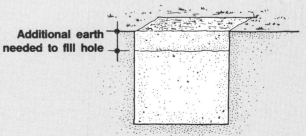

Additional earth needed to fill hole

Compacting for best results

10.   To compact successfully, air and water are expelled from the soil so that its mineral particles can settle very tightly together. For best results, you therefore should always:

- place and compact the soil in **thin horizontal layers** about 15 to 20 cm thick, so air and water can be expelled easily;
- **wet the soil** material to its optimum moisture content for compaction (see Section 102, **Soil**, **6**);
- finish off the slopes of the completed dike to form a well-compacted surface.

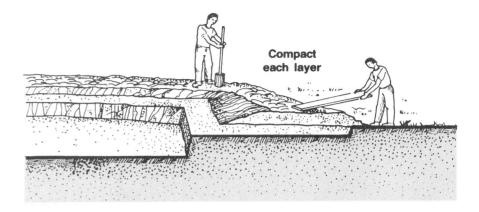

Compact each layer

Note: if the soil to be compacted can be formed into a firm ball that sticks together, the moisture content is adequate for immediate compaction. If the soil is too wet, you should let it dry through evaporation for a while. If the soil is too dry, you should water it slightly and mix well to make it homogeneous.

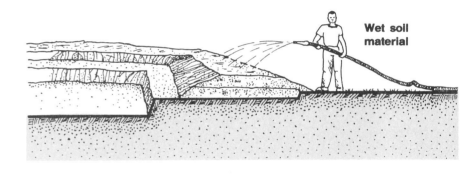

Wet soil material

Good soil material for compacting will stick together

Finish dike slopes

Compacting soil by hand

11. To compact thin layers of soil by hand, you can use simple tools such as:

- a thick stick or the bottom part of a palm frond;
- a thick stick rounded at one end for pounding vertically, for example soil in a trench;
- a hand tamper, a metal or concrete weight (maximum 4 to 6 kg) attached to a wooden handle, with a surface area of about 150 cm^2, which you can make or buy cheaply from hardware stores (see also note on page 260).

12. Hand compacting is generally suitable for small dikes, typically 1 to 1.5 m in height and up to 1 m top width, or smaller if soils are not of good quality.

Note: for clays and similar soils, it may be better to use a kneading action, for example by using the heel of the foot.

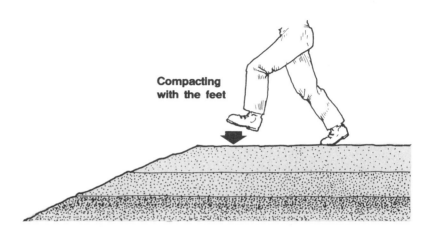

Compacting with the feet

Various compacting tools

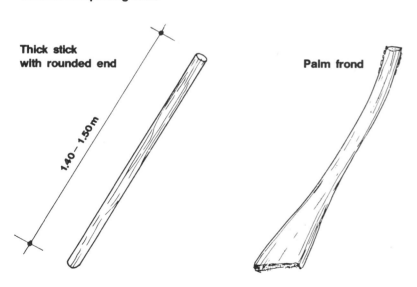

Thick stick with rounded end

1.40 – 1.50 m

Palm frond

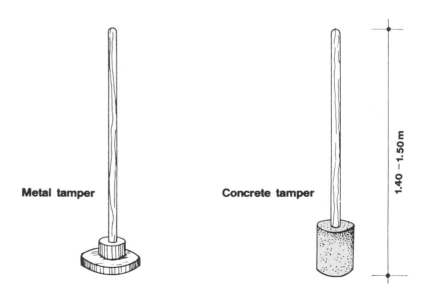

Metal tamper

Concrete tamper

1.40 – 1.50 m

259

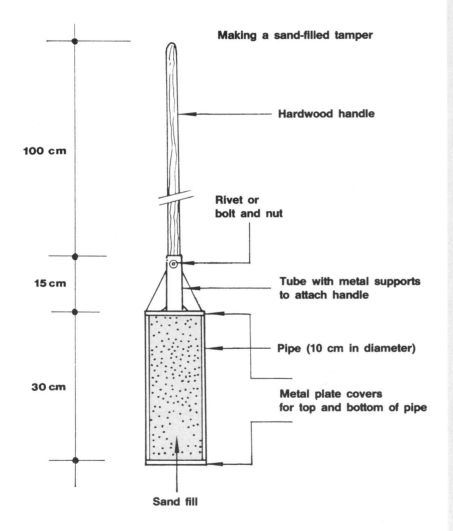

Note: you can easily make a hand tamper
using scrap metal fittings,
a section of pipe filled with sand
and a hardwood handle

Making a sand-filled tamper

Hardwood handle

100 cm

Rivet or
bolt and nut

15 cm

Tube with metal supports
to attach handle

Pipe (10 cm in diameter)

30 cm

Metal plate covers
for top and bottom of pipe

Sand fill

260

Compacting soil with machinery

13. As the dikes and the area to be compacted increase in size, it is better to compact mechanically.

14. For relatively small compaction jobs, you can use **vibration plates** and **percussion tampers** called **frogs**. For bigger jobs, it is usually sufficient to use construction equipment such as tractors and trucks to compact the earth fill by running over it repeatedly. Special compaction equipment such as sheepsfoot rollers, steel-wheel rollers and pneumatic rollers can also be used, if available, under competent supervision. Average output per working hour **(m²/h per 25-cm layers)** for various compactors are as follows:

Compactor	Output m²/h
Percussion tamper (frog)	30-150
Vibration plate	300-600
Roller	
☐ sheepsfoot	1 000
☐ steel-wheel	2 000-5 000
☐ pneumatic	5 000-15 000

Note: the compaction of non-cohesive soils such as sand requires heavy pressure (weight) and, if possible, vibration. On the other hand, the compaction of cohesive soils such as silt and clay requires a kneading action. Thus to compact a clayey soil you cannot use a normal steel-wheel roller which would compact a surface layer only, but you would need a sheepsfoot or pneumatic roller (see Section 102 and **Table 26**, **Soil, 6**).

Compacting small areas

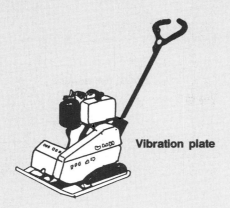

Vibration plate

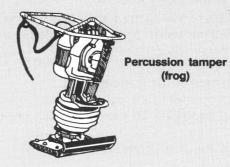

Percussion tamper (frog)

Compacting larger areas

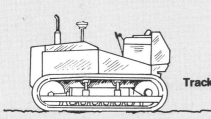

Tracked vehicle

Sheepsfoot roller

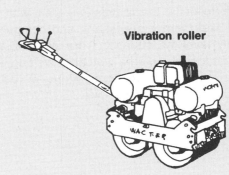

Vibration roller

Pneumatic roller

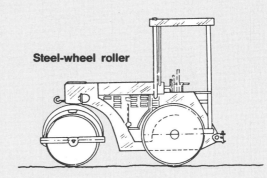

Steel-wheel roller

261

63   Preparing the foundations of the dike

1.   After clearing the site, removing the surface soil and marking out the position of the dike, the foundations of the dike should be prepared. This may include:

- treating the surface of the foundations;
- excavating and backfilling the cut-off trench;
- excavating and backfilling an existing stream channel.

Treating the surface of the foundations

2.   The surface of the dike's foundations needs to be well compacted, so that the dike can be solidly attached without any risk of it sliding away.

(a)   Break the ground surface thoroughly and turn it to a depth of about 15 cm. (You could use a plough or a hoe.)

(b)   Fill all holes in the foundation area with suitable soil. Use thin layers, wet them if necessary, and compact well.

(c)   Roughly level the surface of the dike's foundations.

(d)   Compact the whole area well after wetting, if necessary, so that the surface materials of the foundations are as well compacted as the subsequent layers of the dike.

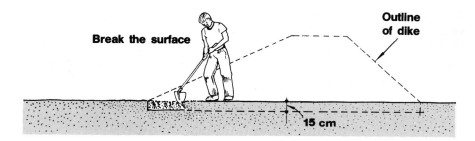

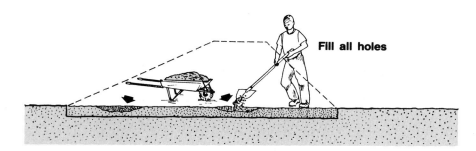

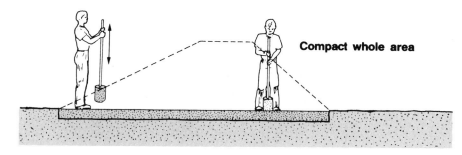

Building a cut-off trench

3. If the foundation soil does not contain an adequate layer of impervious material at the surface, you should build a **cut-off trench** (sometimes called a puddle trench) within the dike's foundations. Its main purpose is to reduce water seepage under the dike. It will also help to anchor the dike solidly to its foundations.

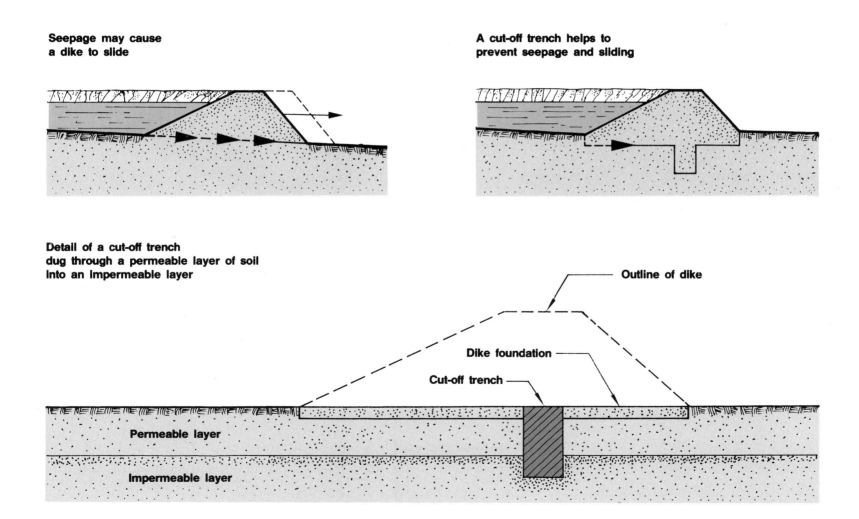

**Seepage may cause
a dike to slide**

**A cut-off trench helps to
prevent seepage and sliding**

**Detail of a cut-off trench
dug through a permeable layer of soil
into an impermeable layer**

Outline of dike

Dike foundation

Cut-off trench

Permeable layer

Impermeable layer

4. The size of the cut-off trench increases with the size of the dike. You can use the following guidelines:

- **width of the trench**: from 0.5 m in small dikes to at least 1 m in larger dikes;
- **depth of the trench**: preferably extending through the pervious soil layer into the impervious layer beneath. In a large dike for a barrage pond for example, the cut-off trench should reach at least 30 cm into the impervious layer, throughout its entire length. In small dikes, the cut-off trench is at the most 0.6 to 1 m deep, regardless of the position of the impervious layer;
- **shape of the trench**: sides are dug vertically in small to medium dikes. In larger dikes, the sides should be cut at a 0.5 to 1:1 slope.

Cut-off trench for large pond

Cut-off trench for small pond

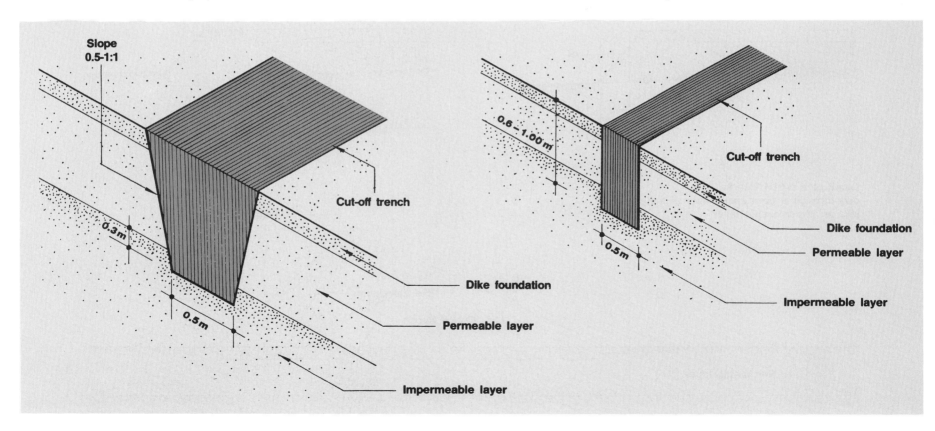

5. To build the cut-off trench, proceed as follows.

(a) Clearly mark the centre line of the dike base, for example with stakes and string.

(b) On each side of this centre line, clearly mark the limit of the cut-off trench to be built.

(c) Dig the trench to the depth, width and side slopes required, placing the removed soil material over the foundations of the dike in one-third of the area toward the dry side of the dike. Be careful not to include roots, organic materials or large stones.

(d) Spread this soil material in thin layers and compact it well.

(e) Make sure the trench is dry.

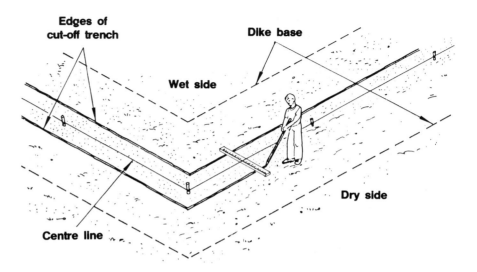

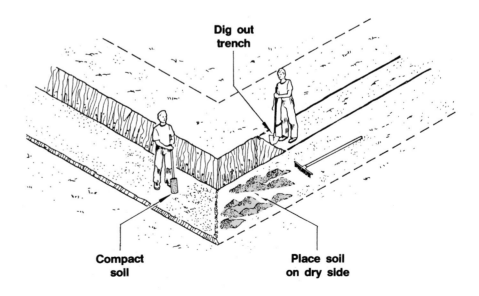

(f) Backfill the cut-off trench to the natural ground surface with soil
material of the same quality as for a dike core (see Section 122,
Soil, **6**). Place the backfill material in thin layers, wet it if necessary,
and compact well. If clay soil is used, "heel" it into place, or use
suitable mechanical equipment.

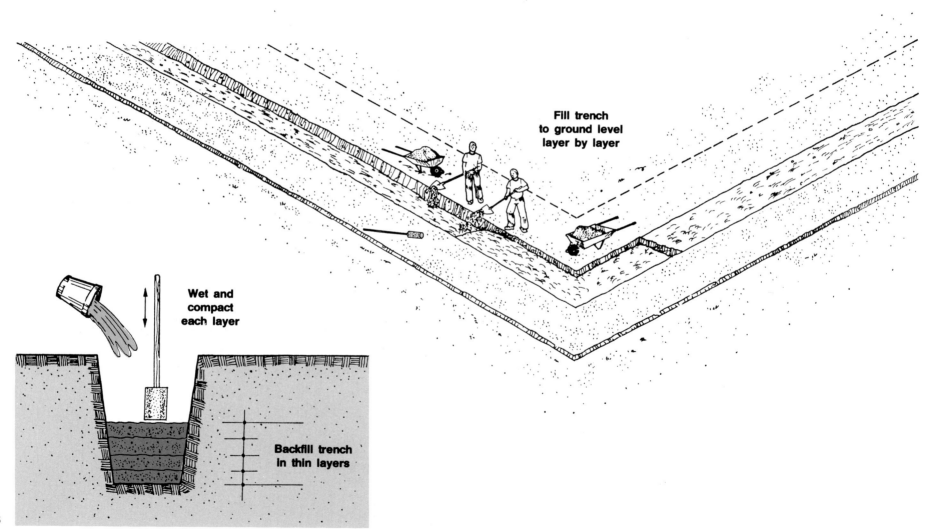

**Fill trench
to ground level
layer by layer**

**Wet and
compact
each layer**

**Backfill trench
in thin layers**

266

Backfilling a stream channel

6. If a stream channel crosses the foundations of the dam, such as in the case of a barrage pond, you must prepare the channel of the stream where the dam will be built. If the water is flowing when you want to work on the channel, you will first have to divert the stream (see pp. 68-79, **Water**, **4**).

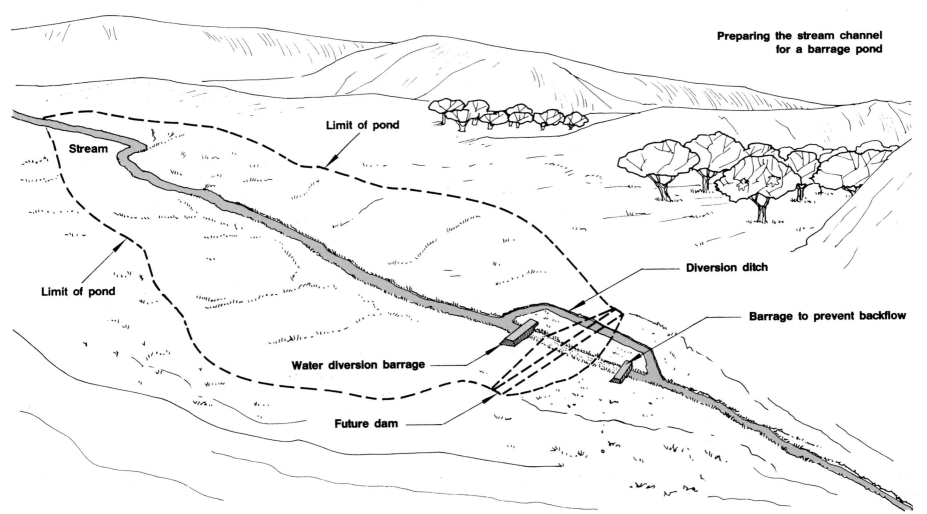

Preparing the stream channel for a barrage pond

Limit of pond

Stream

Limit of pond

Diversion ditch

Barrage to prevent backflow

Water diversion barrage

Future dam

7.   Dig the diversion ditch around the site of the future dam as shown. That way, you will be able to use the same diversion ditch when you build the dam (see Section 66, paragraph 9 and following in this volume). Then, proceed as follows:

(a)   Deepen and widen the channel as necessary to remove all stones, gravel, sand, sediment, stumps, roots and organic matter (see Chapter 5).

(b)   Dig at least 30 cm below the original channel bed or until you reach rock. Make the side slopes of the new channel no steeper than 1:1.

Note: if the soil base below the channel is permeable, it is best to make a cut-off trench.

Cross-sections through a future dam site showing how to clean and enlarge the old stream channel

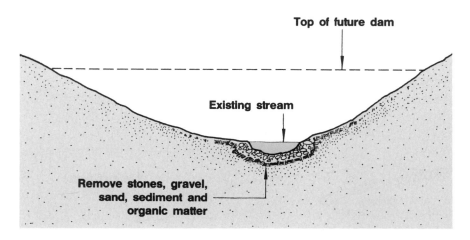

Top of future dam

Existing stream

Remove stones, gravel, sand, sediment and organic matter

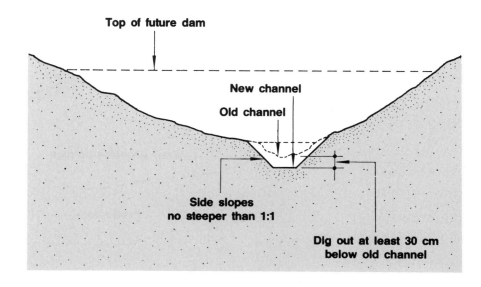

Top of future dam

New channel

Old channel

Side slopes no steeper than 1:1

Dig out at least 30 cm below old channel

64 Calculating dike and excavation volumes

1. Before starting the construction of your pond, you should calculate how much soil you will need to build its dikes. Then, you will need to estimate the excavation volume necessary to provide such soil volume. According to the topography of the construction site and the type of pond to be built, you should select the best method to be used. You should estimate expanded and compacted volumes (see Section 62), and you should also use standard settlement allowances (see **Table 28**).

2. Multiply excavation volume by the expansion factor (**Table 28**) to obtain the expanded volume. This expanded volume is then used in the construction volume of the dike. After compaction and settlement, as estimated by the compaction potential, it should reach the design volume required.

Calculating the width of the dike base

3. Having determined the characteristics of your dikes, determine the width of the dike base (in m) by adding:

- **crest width** (in m);
- construction height (CH in m) multiplied by slope ratio of **dry side** (SD);
- construction height (CH in m) multiplied by slope ratio of **wet side** (SW).

> **Base width = crest width + (CH × SD) + (CH × SW)**

Note: use the construction height, including the settlement allowance, and not the design height of the dike (see Section 61).

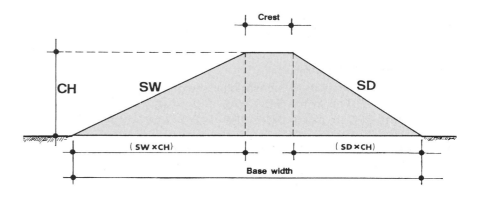

Example

A 0.04-ha pond (400 m²) has to be built in clayey soil with dikes 1.50 m high and 1 m wide at the top, according to the design. If SD = 1.5:1 and SW = 2:1, calculate the base width of the dikes.

(a) From **Table 28**, estimate the settlement allowance of the expanded clay volume (20 percent for medium clay soils).
(b) Consider the design height = (100% − 20%) = 80 percent of construction height.
(c) Obtain the construction height = 1.50 m ÷ 0.80 = 1.88 m.
(d) Calculate dike base width = 1 m + (1.88 m × 1.5) + (1.88 m × 2) = 1 m + 2.82 m + 3.76 m = 7.58 m.

Note: see also examples in **Table 27**.

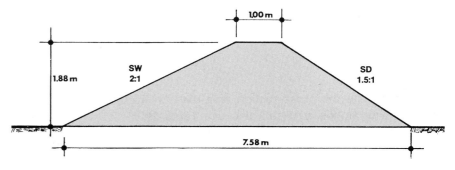

Calculating the cross-section of a dike on horizontal ground

4.   The size of the **cross-section of a dike** on horizontal ground (ABCD in m²) (see diagram opposite) is obtained by adding:

- area ABFE (in m²) =
  crest width (AB) × construction height (CH);
- area AED (in m²) = ED × (AE ÷ 2) = (SD × CH) × (CH ÷ 2);
- area BFC (in m²) = FC × (BF ÷ 2) = (SW × CH) × (CH ÷ 2).

CH = the construction height of the dike;
SD = the slope ratio of the dry side;
SW = the slope ratio of the wet side.

(See **Topography, 16/2**.)

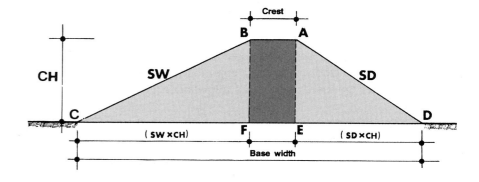

Example

For the above 0.04-ha pond to be built in clayey soil, calculate the size of the cross-section of the dike as:

- area 1 = 1 m × 1.88 m = 1.88 m²;
- area 2 = (1.5 × 1.88 m) × (1.88 m ÷ 2) = 2.6508 m²;
- area 3 = (2 × 1.88 m) × (1.88 m ÷ 2) = 3.5344 m²;
- cross-section = 1.88 m² + 2.6508 m² + 3.5344 m²
  = 8.0652 m².

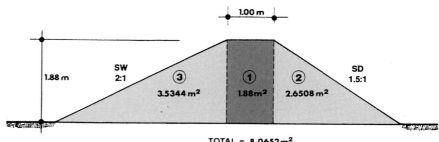

TOTAL = 8.0652 m²

5.   To calculate the cross-section of a dike on horizontal ground **with identical side slopes**, you can also use **Table 29**.

TABLE 29

Cross-sections of dikes on horizontal ground with identical side slopes
(in m²)

Construction height of dike (m)	Side slopes 1.5:1			Side slopes 2:1		
	Top width			Top width		
	1 m	2 m	3 m	1 m	2 m	3 m
0.5	0.8	1.3	1.8	1.0	1.5	2.0
1.0	2.5	3.5	4.5	3.0	4.0	5.0
1.5	5.0	6.5	8.0	6.0	7.5	9.0
2.0	8.0	10.0	12.0	10.0	12.0	14.0
2.5	12.0	14.5	17.0	15.0	17.5	20.0
3.0	16.5	19.5	22.5	21.0	24.0	27.0

Calculating the cross-section of a dike on sloping ground

6.   The cross-section of a dike **on sloping ground** can be calculated most easily using a scale drawing.

(a)   Draw a horizontal line from D, meeting AE at E′.

(b)   Draw a horizontal line from C, meeting BF at F′.

(c)   Draw a vertical line PO down the centre line of the dike.

(d)   Cross-section = ADE + AEFB + BFC = 0.5(AE × DE′) + (AB × PO) + 0.5(BF × F′C).

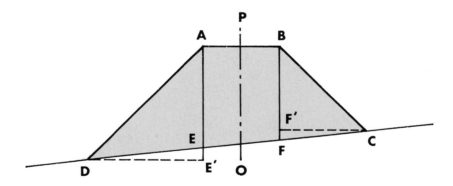

Calculating the cross-section of a dike on sloping ground using a scale drawing

Note: for ground slopes of less than 10 percent, and where the dike side slopes are the same on each side, you can use the method given for horizontal ground.

Calculating the cross-section of a dike on irregular ground

7. The cross-section of a dike to be built **on irregular ground** can be
calculated in two ways.

(a) Draw a straight line D'E'F'C', approximating the shape of the
     ground, then use the procedure given for sloping ground.
(b) Alternatively, mark the shape out on squared paper, and using the
     scale, count the squares to obtain the area (see Section 103,
     Topography, 16/2).

**Calculating the cross-section
of a dike on irregular ground
using a scale drawing**

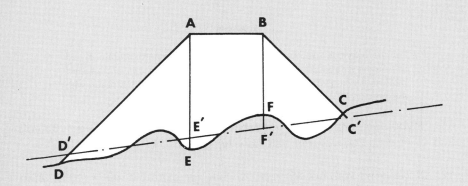

**Calculating the cross-section
of a dike on irregular ground
using squared paper**

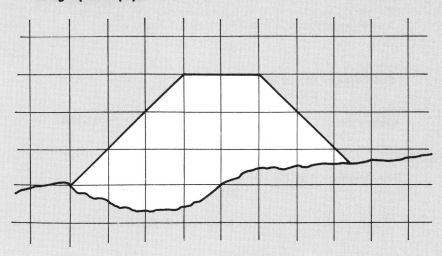

1 cm = 0.5 m
1 square of 0.5 m × 0.5 m = 0.25 m²
15.2 squares × 0.25 m² = 3.8 m²

8. To estimate how much soil will be needed for the construction of a dike, you need to know its volume. The calculation method depends on the site topography and on the type of pond to be built.

9. If **the topography** of the construction **site is reasonably flat** (less than 0.30 m difference in average site levels) and regular, you can calculate the volume of the dike (in m³) by multiplying the **cross-section of the dike** (in m² and halfway along the dike for an average area) **by its length** measured along the centre line (in m).

Example

Using the figures from the example on page 270, the cross-section of the dike equals 8.0652 m². If the length of the dike to be built is 20 m × 4 = 80 m, its volume is 8.0652 m² × 80 m = 653.216 m³.

Pond area = 20 × 20 m
= **400 m²**

Crest width = **1 m**
SW = **2:1**
SD = **1.5:1**

Dike height = **1.88 m**

→ **Crest**

Centre line of dike crest

10. Alternatively you can **calculate the volume using graphs**.

(a) In **Graph 3a**, enter the area of the pond (in m²). According to the average construction height of the dikes (in m), find the **standard volume** (in m³) of the dikes for a **standard pond** where: ratio length:width is 1:1 (square shape); both dike slope ratios are 2:1; crest width is 1 m.

(b) If the side slopes of the dike are not 2:1, multiply the standard volume by **S**, according to the table shown in the next column.

Inner slope	Outer slope	S
1	1	0.63
1.5	1	0.72
1.5	1.5	0.82
2	1.5	0.90
2	2	1.00
2.5	2	1.09
2.5	2.5	1.18
3.0	2.5	1.27
3.0	3.0	1.36
3.5	3.0	1.46
3.5	3.5	1.55
4.0	3.5	1.65

(c) If the crest width of the dike is not 1 m, multiply the standard volume by **C**, obtained from **Graph 3b**.

(d) If the shape of your pond is not square, multiply the standard volume by **P**, obtained from **Graph 3c**.

Example

For the previous example, **Graph 3a** shows a standard volume of 720 m³. As the side slopes are 2:1 (inner) and 1.5:1 (outer), this is multiplied by S = 0.9 to give 720 m³ × 0.9 = 648 m³. (Compare this result with the previous example in which you calculated 653 m³.)

11. If you decide to change the **crest width** to 0.51 m, from **Graph 3b** you find **C** = 0.8. The volume will now be 648 m³ × 0.8 = 518.4 m³.

12. If the pond was not 20 × 20 m but, for example, 40 × 10 m, then the ratio L:W = 4. From **Graph 3c**, **P** = 1.25. With a crest width of 1 m, the volume of the dikes would then be 648 m³ × 1.25 = 810 m³.

Pond area = 10 × 40 m
= **400 m²**

Crest width = **1 m**
SW = **2:1**
SD = **1.5:1**

Dike height = **1.88 m**

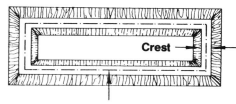

Crest →

Centre line of dike crest

Dike volume for a standard square pond
(crest width = 1 m; dike slope ratios 2:1)

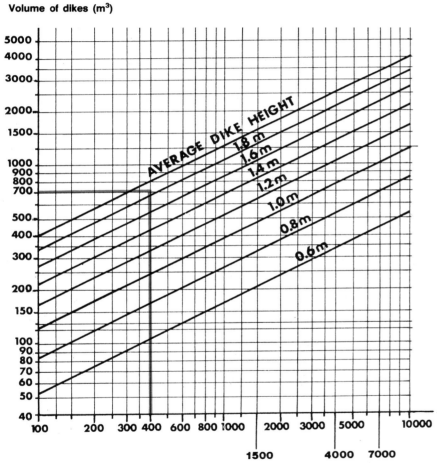

Volume of dikes (m³)

AVERAGE DIKE HEIGHT

1.8 m
1.6 m
1.4 m
1.2 m
1.0 m
0.8 m
0.6 m

Area of pond (m²)

Crest width correction factor

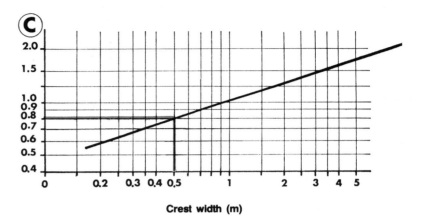

C

Crest width (m)

Pond shape correction factor

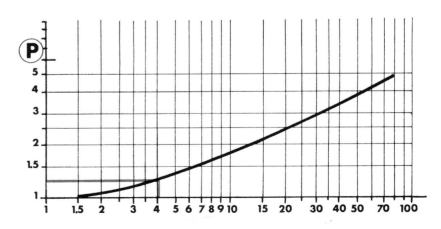

P

Pond length: width ratio

275

Calculating the volume of dikes on sloping or irregular ground

13. If the topography of the site is more steeply sloping or more irregular, you cannot calculate the volume of the pond dikes just by using one cross-section. There are several possible methods, depending on the type of ground and the accuracy you require.

14. With a first group of methods you can calculate the dike volumes by using **averages of the dike cross-sections** or you could use the average of the **cross-sections at the corners** of the dike.

Average of areas at corners of dike

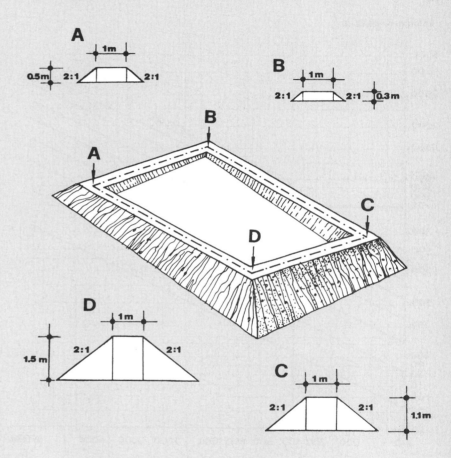

Example

A 400-m² (20 × 20 m) pond is to be constructed with wall heights of 0.5 m at corner A, 0.3 m at corner B, 1.1 m at corner C and 1.5 m at corner D. Crest width is 1 m and side slope 2:1 on both sides. The cross-section areas at each corner are:

A: (1 m × 0.5 m) + 2 × (0.5 m × 0.5 m × 1 m) = 1.5 m²,
B: (1 m × 0.3 m) + 2 × (0.5 m × 0.3 m × 0.6 m) = 0.48 m²,
C: (1 m × 1.1 m) + 2 × (0.5 m × 1.1 m × 2.2 m) = 3.52 m²,
D: (1 m × 1.5 m) + 2 × (0.5 m × 1.5 m × 3 m) = 6.0 m².

Average area for wall AB = (1.5 m² + 0.48 m²) ÷ 2 = 0.99 m² and volume for wall AB = 0.99 m² × 20 m = **19.8 m³**.

Similarly:

- for BC, average area = 2 m² and volume = **40 m³**;
- for CD, average area = 4.76 m² and volume = **95.2 m³**;
- for DA, average area = 3.75 m² and volume = **75 m³**.

Consequently, total volume of dikes = 19.8 m³ + 40 m³ + 95.2 m³ + 75 m³ = 230 m³.

15. Alternatively with rough ground, you could use average dike cross-sections based on an **estimated base line**, then add the four wall volumes.

Example

Using the example on page 276, the heights of A and D are estimated by drawing line XY through the base so that the areas above the line are approximately equal to those below the line. Note that the profile of the ground as drawn should represent the average height across the wall base.

Estimating dike base on rough ground

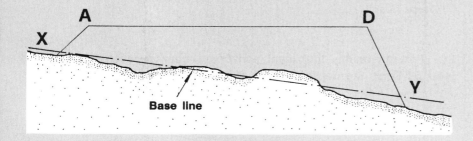

16. You can also use the **graphical method** explained earlier (see paragraph 10 on page 274), using an average height for the four walls of the dike, although this method is less accurate.

Example

Using the graphical method, the average wall height is (0.5 m + 0.3 m + 1.1 m + 1.5 m) ÷ 4 = 0.85 m, and the standard volume, which needs no further correction, is about 180 m³. This is about 80 percent of the previous figure (see paragraph 14 on page 276).

17. For a more accurate measurement of dike volume on rough ground, you should apply the following formula, known as **Simpson's rule**, where:

$$V = (d \div 3) \times [A1 + An + 4(A2 + A4 + \ldots An\text{-}1) + 2(A3 + A5 + \ldots An\text{-}2)].$$

(a) Divide the length of the dike into an **odd number n** of cross-sections at equal intervals of **d** metres.
(b) Calculate the area **A** of each cross-section as explained earlier.
(c) Introduce these values into the above formula.

Example

The dike is 60 m long.

(a) At intervals **d** = 10 m, identify seven cross-sections A1 ... A7 and calculate their respective areas to obtain A1 = 10 m²; A2 = 16 m²; A3 = 18 m²; A4 = 11 m²; A5 = 8 m²; A6 = 10 m²; A7 = 12 m².
(b) Introduce these values into the Simpson's rule formula:

$$V = (d \div 3) [A1 + A7 + 4(A2 + A4 + A6) + 2(A3 + A5)].$$

(c) Calculate $V = (10\ m \div 3) [10\ m^2 + 12\ m^2 + 4(16\ m^2 + 11\ m^2 + 10\ m^2) + 2(18\ m^2 + 8\ m^2)] = 740\ m^3.$

Calculating dike area by cross-section

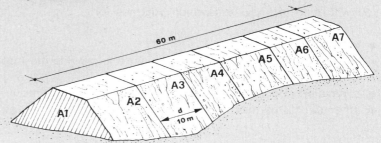

Calculating the volume of a dam for a barrage pond

18. If you have to calculate the volume of the dam to be built for a barrage pond, one of the above methods can be applied. However, because of the presence of the stream channel and numerous changes of ground slopes, it is usually necessary for precise estimates either to measure cross-sections at small **d** intervals or to subdivide the dam into sections using different **d** intervals. (For a more rapid but less precise estimate, see Section 113, **Topography**, **16/2**.)

**Cross-sections to be calculated
for a barrage dam**

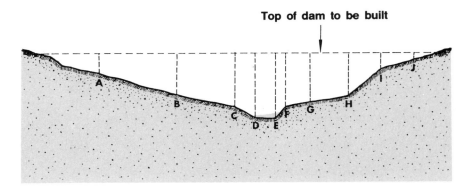

20. You will normally have to remove the topsoil before you reach soil good for construction material. Levels should therefore be taken **from the base of the topsoil** layer. In most cases, the sides of the excavation should be sloped to prevent them from collapsing. In many cases (ponds, channels, etc.) these will be of specified gradients.

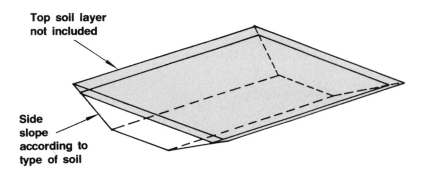

21. For reasonably flat, level surfaces, where **excavated width is at least 30 times the depth**, volume of excavation can be estimated as:

V = top area × depth of excavation.

Calculating volumes of excavated material

19. You will need to know excavation volumes for:

- topsoil;
- borrow pits, dug near an earth structure to provide the material for its construction;
- excavated ponds, to provide the pond volume required;
- other structures such as harvest pits, supply channels, etc.

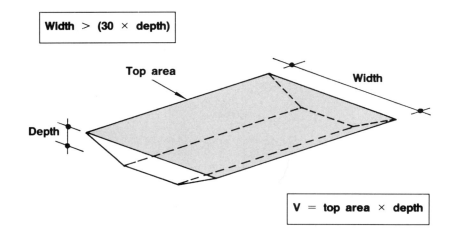

22. Where the **width is less than 30 times the depth**, you should correct for side slopes as follows:

$$V = [(\text{top area} + \text{bottom area}) \div 2] \times \text{depth}.$$

Example

A 400-m^2 (40 × 10 m) area is to be excavated, 1 m deep, with side slopes 2:1. As the width (10 m) is less than 30 times the depth (30 × 1 m), the first method is not accurate (estimated volume would be 400 m^2 × 1 m = 400 m^3).

Use the second method, where top area = 400 m^2 and base area = base length × base width.
Base length = 40 − (2 × slope × depth) = 40 − (2 × 2 × 1 m) = 36 m
Base width = 10 − (2 × slope × depth) = 10 − (2 × 2 × 1 m) = 6 m
Base area = 36 m × 6 m = 216 m^2
Average area = (400 m^2 + 216 m^2) ÷ 2 = 308 m^2
Volume therefore = 308 m^2 × 1 m = 308 m^3.

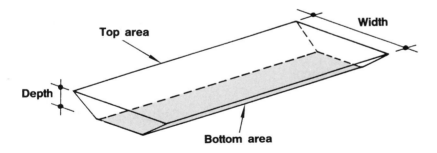

Width ≤ **(30 × depth)**

Top area

Width

Depth

Bottom area

$$V = [(\text{top area} + \text{bottom area})] \div 2 \times \text{depth}$$

23. **On gently sloping ground**, calculate the cross-section at each end of the excavation. Then:

(a) Calculate the average cross-section of the excavation.
(b) Multiply by the average length of the excavation.

Example

If the area is on a gentle slope, calculate cross-sections at AB and CD, and average length.

Calculating volume of excavation by cross-sections

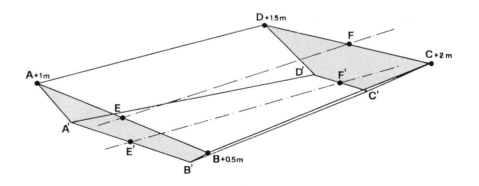

This example is continued on page 280.

Example, continued

(a) Cross-section at AB can be determined by drawing out on squared paper, or estimated by: [(AB + A′ B′) ÷ 2] × average depth, or [(10 + 7) ÷ 2] × [(1 + 0.5) ÷ 2] = 8.5 m × 0.75 m = 6.375 m².

Cross-section at AB

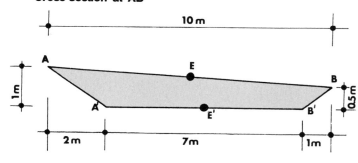

(b) Cross-section at CD, similarly = [(10 + 3) ÷ 2] × [(2 + 1.5) ÷ 2]
= 6.5 m × 1.75 m = 11.375 m².

Cross-section at CD

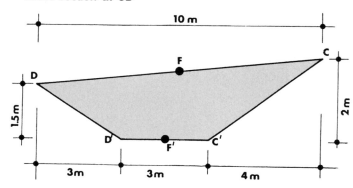

(c) Average length can be determined at the midpoint. Average length = (top length EF + bottom length E′ F′) ÷ 2 = (40 m + 35 m) ÷ 2 = 37.5 m.

(d) Thus volume = average area × average length = [(6.375 m² + 11.375 m²) ÷ 2] × 37.5 m = 332.8 m³.

24. **On more steeply sloping ground** (steeper than 10 percent in any direction), you can use the same method as above; however, the lengths of the base and the corresponding cross-sections, as calculated in the previous method, will not be sufficiently accurate. To obtain a reasonable estimate proceed in the following way.

(a) Use squared paper and obtain the **base length** by measurement. Then use this base length in the calculations, as shown earlier.

Finding base length by measurement

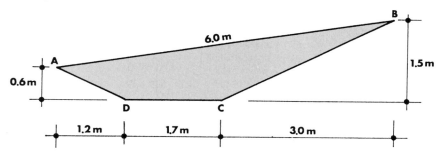

(b) For best accuracy, calculate **cross-section ABCD** = area ADC + area ABC = [(FC × AF) ÷ 2] + [(EC × AB) ÷ 2]

Cross-section ABCD

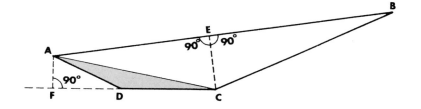

25. With particularly **uneven surfaces**, you can use one of the following methods.

- Estimate the surface level by averaging the elevations of specific points of the surface, and then calculating the cross-sections as above.
- For a more accurate result, use **Simpson's rule** with a series of cross-sections (see paragraph 16 in this section).
- Set up a **grid over the area** and calculate the volume (in m³) either section by section (see **Topography, 16/1**), or by noting the elevation of each grid crossing point (in m) and using the formula:

---

volume = [(**A** ÷ **4**) × (**sum of the elevation of single points**)]
      + [**2** × (**sum of double points**)]
      + [**3** × (**sum of triple points**)]
      + [**4** × (**sum of quadruple points**)]

---

where **A** is the area of each grid square in m².

Example

In the case shown, relative elevations are marked on a grid made of 10 × 10 m squares so that the area of grid squares A = 10 × 10 m = 100 m². According to the formula:

Volume = (100 m² ÷ 4) × [(3.1 m + 2.0 + 2.6 + 2.0 + 3.1) + 2(2.6 m + 3.5+3.0+2.0+3.5+2.5+1.8+ 2.0)+3(2.8 m)+4(3.1 m+2.1+2.5)]
= (100 m² ÷ 4) × [(12.8 m) + 2(20.9 m) + 3(2.8 m) + 4(7.7 m)]
= (100 m² ÷ 4) × (93.8 m) = 2 345 m³.

Note: you will normally have to **correct this volume for the side slopes**. It is usually easier to make these adjustments outside the grid, calculating the additional volume either square by square or by averaging along each side of the grid.

Relative elevations

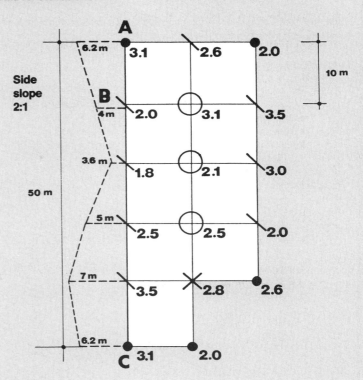

KEY

● Points used in only one square (single points)

＼ Points used in two squares (double points)

✕ Point used in three squares (triple point)

○ Points used in four squares (quadruple points)

281

If in the above example, a side slope ratio of 2:1 is used, the additional volume can be estimated in two ways.

(a)   Estimating square by square:

At the first square (Section AB) for example:

- average height = (3.1 m + 2.0 m) ÷ 2 = 2.55 m;
- average width = (6.2 m + 4.0 m) ÷ 2 = 5.10 m;
- volume = 0.5 (height × width) × length = 0.5(2.55 m × 5.10 m) × 10 m = 65 m³.

(b)   Estimating by averaging along each side.

For side AC for example:

- average height = (3.1 m + 2.0 m + 1.8 m + 2.5 m + 3.5 m + 3.1 m) ÷ 6 = 2.66 m;
- average width = (6.2 m + 4 m + 3.6 m + 5 m + 7 m + 6.2 m) ÷ 6 = 5.33 m;
- volume = 0.5(2.66 m × 5.33 m) × 50 m = 354.4 m³.

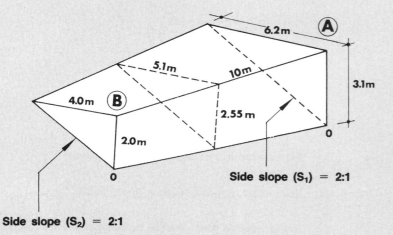

Side slope (S_2) = 2:1

26. To estimate the **volume at each corner**, use the formula:

$$V = 0.33 \times h \times S1h \times S2h,$$

where **h** = depth of excavation (in m) at the corner and S1, S2 are the side slopes.

In the previous case, at corner A for example, if side slopes on both side and end are S1 = S2 = 2:1, volume of corner cut = 0.33 × 3.1 m × (2 × 3.1 m) × (2 × 3.1 m) = 39.7 m³.

If the end slope had been 3:1 and the side slope 2:1, then volume = 0.33 × 3.1 m × (3 × 3.1 m) × (2 × 3.1 m) = 59.6 m³.

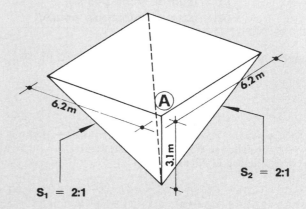

Remember: with any calculations for building and excavation, do not use methods that are more precise than you require. Because of the difficulty of predicting expansion and compaction accurately, volume estimates in practice are generally only accurate to within 10 percent. There is usually therefore little point in being more precise than this, and so there is no need to allow for every little irregularity or slight change in slope.

1. Dug-out ponds, entirely obtained through soil excavation, are the simplest to build. There are two main types of dug-out pond depending on the water supply (see Section 14):

- dug-out ponds fed by **rain and surface runoff**, commonly found in relatively flat, well-drained terrain such as the low point of a natural depression;
- dug-out ponds fed by **springs or seepage**, the latter being commonly found in areas where the ground water table is close to the surface, either permanently or seasonally.

**An example
of seepage dug-out ponds**

**Note: see also page 38 for a larger illustration
of seepage dug-out ponds in a valley bottom
and page 285 for an illustration of this site
before pond construction**

Kinds of dug-out ponds

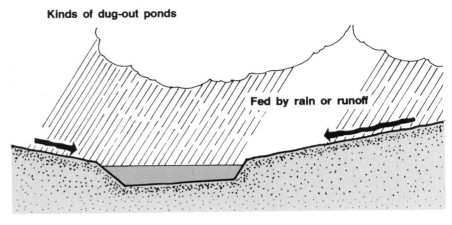

Fed by rain or runoff

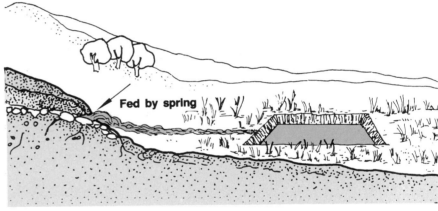

Fed by spring

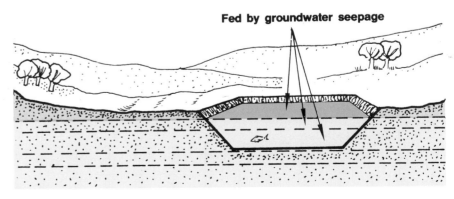

Fed by groundwater seepage

283

2. To build a **rain-fed dug-out pond**, it is essential to have enough impervious soil at the site to avoid excess seepage losses. The best sites are those where fine-textured clays and silty clays extend well below the proposed pond depth. Sandy clays extending to adequate depths are satisfactory. Avoid sites with porous soils, either at the surface or at the depths through which the pond would be cut.

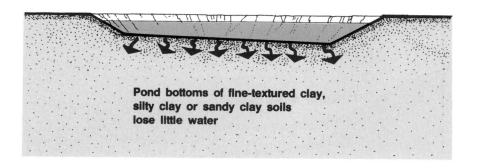

Pond bottoms of fine-textured clay, silty clay or sandy clay soils lose little water

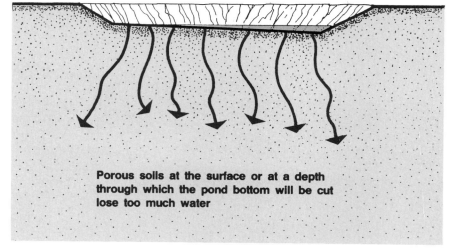

Porous soils at the surface or at a depth through which the pond bottom will be cut lose too much water

3. To build a **seepage dug-out pond**, look for soils where the water-bearing layer is thick enough and permeable enough to provide the required water. It is best to observe the site during a complete annual cycle to check on the possible variations of the water table elevation with the season.

Site before construction

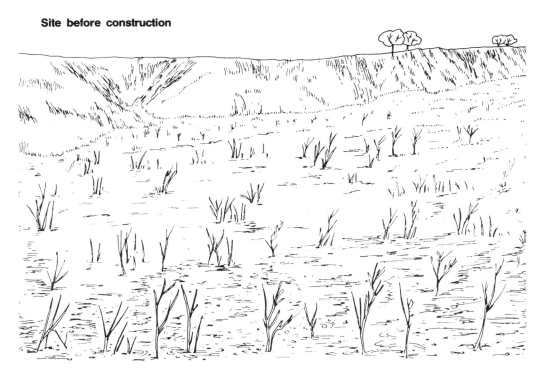

Check water table variations on the site for one year

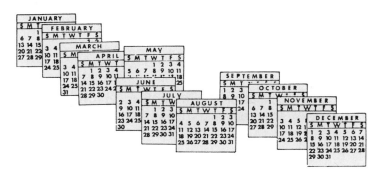

Building a dug-out pond

4. Begin building the dug-out pond by **preparing the site** as follows.

(a) **Mark the area to be cleared** using stakes. This should include the total area of the pond to the outside limits of the pond dikes and, in addition, an area of two to three metres to serve as a work space and for walkways beyond the dikes.

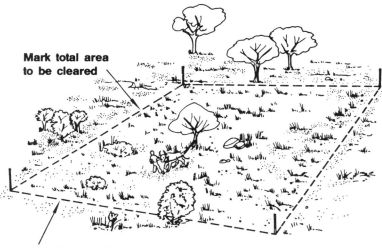

Mark total area to be cleared

Area includes 2 to 3 m for work space and walkways beyond the pond dikes

(b) **Clear all vegetation** from the marked area (see Chapter 5). Also remove all shrubs and trees within 10 metres of the cleared area.

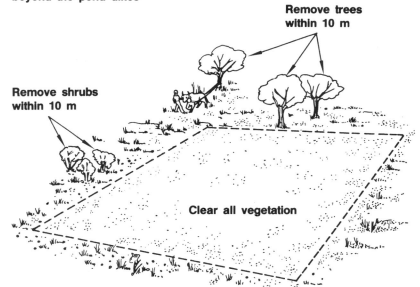

Remove trees within 10 m

Remove shrubs within 10 m

Clear all vegetation

(c) Next, at the very centre of the cleared area, mark the pond area to the **outside limits of the pond dikes** using heavy string or cord. Remove the surface soil from this area and store it for later use.

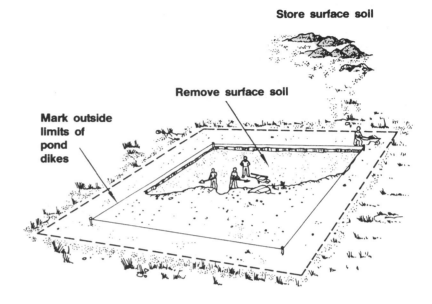

Store surface soil

Remove surface soil

Mark outside limits of pond dikes

(d) Now mark the **inside limits of the pond bottom** using heavy string or cord. Do this on the basis of the selected side slopes (see Item 13, Section 61).

(e) When staking out the pond bottom, indicate on each stake the depth of excavation from ground surface to pond bottom (see Item 10, Section 114, **Topography**, **16/2**).

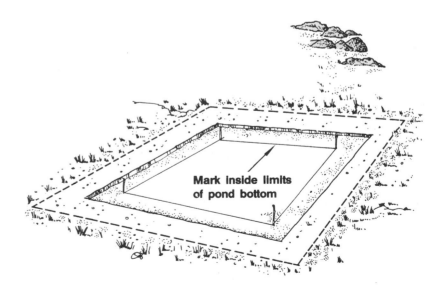

Mark inside limits of pond bottom

(f)   There are **two easy ways to dispose of waste soil material** (see also illustration at top of opposite page) and to prevent it from tumbling or eroding back into the dug-out pond:

● if there is enough space around the pond, you can **spread the waste soil** there. Limit the thickness of spread soil to 1 m at most and slope it gently away from the pond;

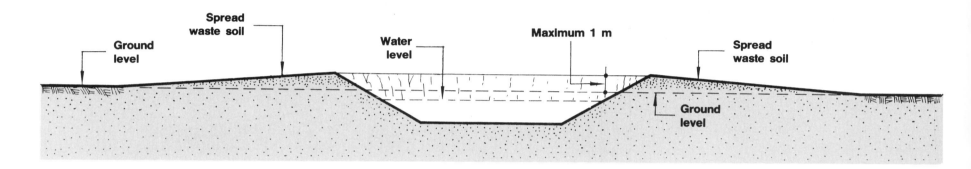

● make a **pile of waste soil** near the pond, but be sure to leave at least 4 m between the toe of the pile and the pond. The sides of piled soil should have a gentle slope of 3:1 or more.

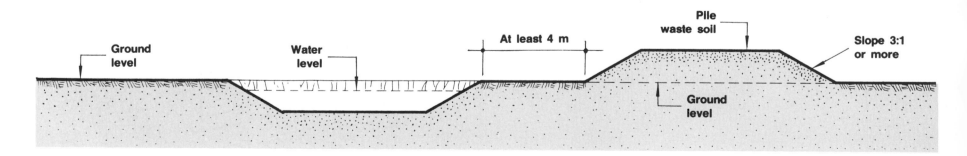

Note: you can use a pile of soil as a windbreak or for the cultivation of a crop (see also Section 56).

(g) Clearly mark the limits of the areas where the excavated material will be spread or piled.

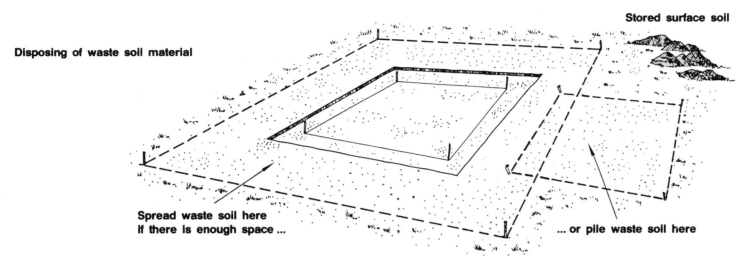

Disposing of waste soil material

Stored surface soil

Spread waste soil here if there is enough space ...

... or pile waste soil here

(h) Dig to the designed depth within the limits of the pond, **cutting the sides vertically**. Transport the waste soil to the planned areas.

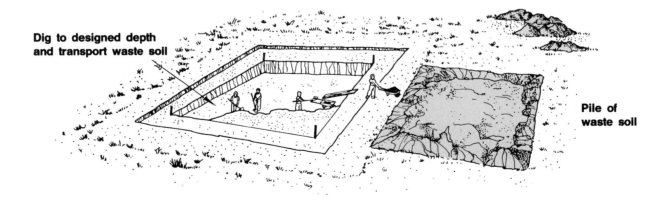

Dig to designed depth and transport waste soil

Pile of waste soil

Note: usually the pond bottom in drainable ponds is given a one-percent slope between the inlet and outlet ends; in undrainable ponds, the **pond bottom may be horizontal**. You can calculate the volume of the material to be excavated using one of the methods in the previous section.

(i)   **Shape the sides** of the pond to the desired slope and finish the pond
      bottom and the horizontal dike tops. Remove any excess soil.

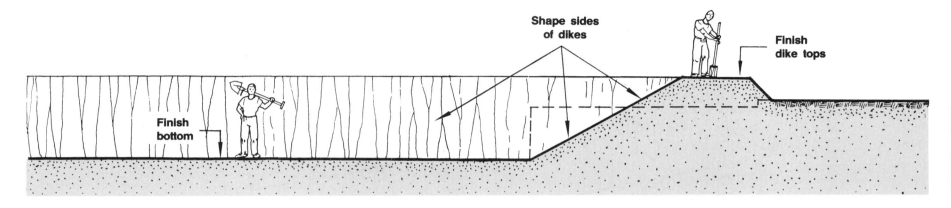

(j)   **Bring back the surface soil** to cover the waste material and the dike
      tops. Then plant or sow grass all around the pond to prevent
      erosion (see Section 69).

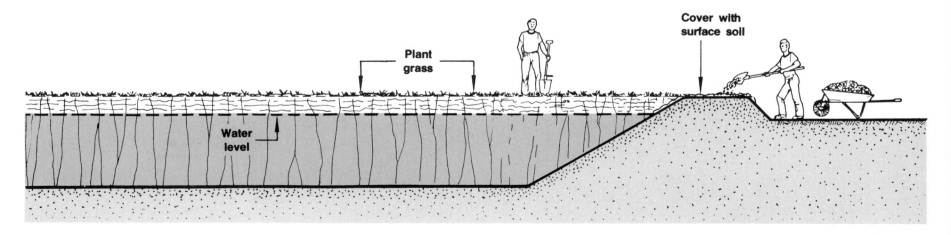

Note: water control structures such as a feeding canal, inlet pipe, outlet
sluice, spillway or drainage canal may be provided for dug-out ponds
(see **Pond construction, 20/2**).

66 Constructing barrage ponds

1. Barrage ponds are embankment ponds formed by a **dam**, which is built across a narrow valley to impound water behind it (see Section 14).

Note: in this manual, for barrage ponds you will learn how to build small dams only, their height being limited to 2.50 m. If you need to build a higher dam, you should consult a specialized engineer.

2. As the height of the dam increases, the need for **sound foundations** becomes essential. The best foundations consist of a thick layer of relatively impermeable consolidated clay or sandy clay, at shallow depth. Never build a dam over rock or sand. Whenever in doubt, ask for advice.

Obtaining the soil material for construction

3. You can calculate the volume of material required using one of the methods described earlier (see Section 64, paragraph 18).

4. In order to reduce the distance the soil needs to be transported, try to dig the soil necessary for the construction of the dam **from an area close by**, for example:

- from the edges of the valley;
- from inside the pond.

5. An area from which soil is taken is called a borrow pit. Be careful to keep the limit of any borrow pit at least 10 m from the wet toeline of the dam. Plan the drainage of this area to be included within the barrage pond, for example using a trench toward the water outlet.

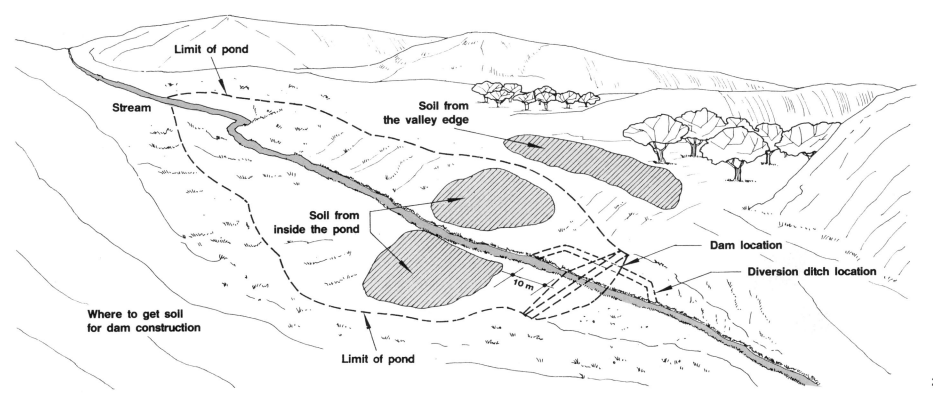

6.   Clearly mark the **centre line of the dam** at ground level with tall stakes and a line. It is usually perpendicular to the main axis of the stream in the valley to be flooded.

7.   Calculate **distances** from the **centre line to the two toelines** on a series of perpendiculars set out at regular intervals as: (dam crest width ÷ 2) + (dam construction height × slope ratio).

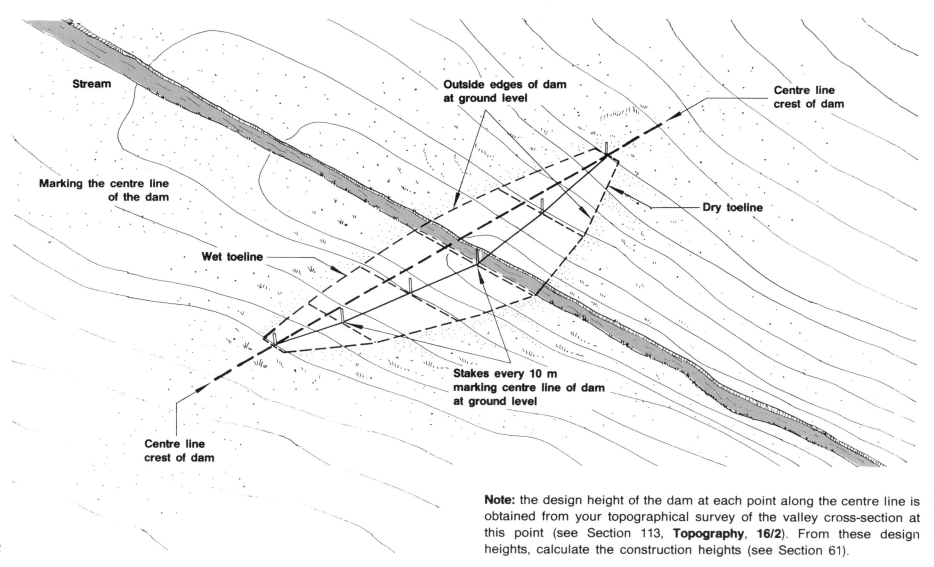

Stream

Marking the centre line
of the dam

Wet toeline

Centre line
crest of dam

Outside edges of dam
at ground level

Centre line
crest of dam

Dry toeline

Stakes every 10 m
marking centre line of dam
at ground level

Note: the design height of the dam at each point along the centre line is obtained from your topographical survey of the valley cross-section at this point (see Section 113, **Topography**, **16/2**). From these design heights, calculate the construction heights (see Section 61).

Example

You plan to build a dam with maximum design height **DH** = 2.10 m, crest width = 2 m, wet slope 2:1 and dry slope 1.5:1.

Soil settlement allowance is estimated at 15 percent. The valley cross-section along the centre line of the dam can be sketched as shown, giving the design heights DH(A), DH(B) ... at points A, B ... at 10-m intervals along the centre line.

Calculate the distances from centre line AF to toelines GHIK and LMNO as follows:

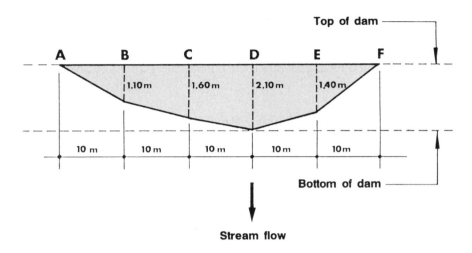

Top of dam

1.10 m	1.60 m	2.10 m	1.40 m	

| 10 m | 10 m | 10 m | 10 m | 10 m |

Bottom of dam

Stream flow

Point	Slope ratio	Dam design height DH (m)	Dam construction height CH (m)*	Distance from centre line to toelines (m)**
B	2:1	1.10	1.29	BG = 3.58
C	(wet)	1.60	1.88	CH = 4.76
D		2.10	2.47	DI = 5.94
E		1.40	1.65	EK = 4.30
B	1.5:1	1.10	1.29	BL = 2.94
C	(dry)	1.60	1.88	CM = 3.82
D		2.10	2.47	DN = 4.71
E		1.40	1.65	EO = 3.48

* Calculate CH = DH ÷ [(100 − SA) ÷ 100] where SA is the settlement allowance in percent;
here SA = 15 percent and CH = DH ÷ 0.85
** As (crest width ÷ 2) + (CH × slope)

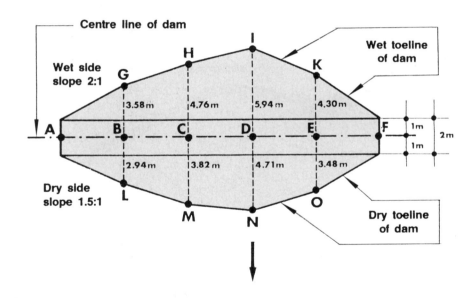

Centre line of dam

Wet toeline of dam

Wet side slope 2:1

| 3.58 m | 4.76 m | 5.94 m | 4.30 m |

| 2.94 m | 3.82 m | 4.71 m | 3.48 m |

Dry side slope 1.5:1

Dry toeline of dam

8. Stake out on the ground points G, H, I, K on the wet side and L, M, N, O on the dry side of centre line AF. These points show you what the outside limits of the dam base should be.

9. **Divert the stream** to a site as close as possible to one of the valley sides and well away from the original stream bed (see page 267, earlier in this chapter, and also pages 68 and 69, **Water**, **4**). This task will be much easier if you have scheduled the construction work to coincide with the dry season.

10. To **prepare the foundation** of the dam, clear the base area, remove the surface soil and treat the surface of the foundations, giving particular attention to the old stream channel (see Section 63) and to the sides of the valley, according to the quality of the foundations' soil:

11. **If the soil is impermeable**, dig an anchoring trench (about 1 m wide and 0.4 m deep) along the centre line of the dike base to anchor the dike to its foundations. Refill this trench with good clayey soil and compact it well. Extend the trench sideways, well into the sides of the valley;

12. **If the soil is permeable**, build a cut-off trench (at least 1.5 m wide) along the centre line of the dam (see Section 63), which will also assist in anchoring the dam to its foundations. Extend the trench sideways, well into the sides of the valley.

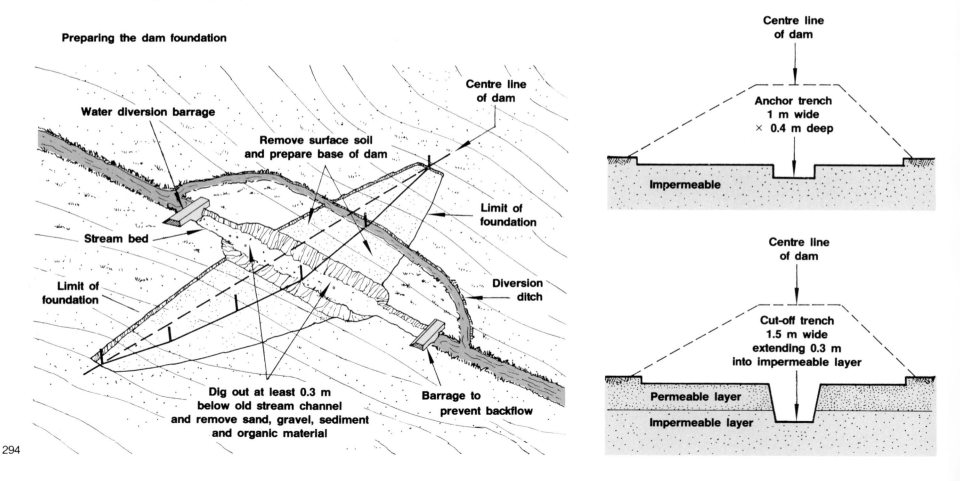

Preparing the dam foundation

Water diversion barrage

Remove surface soil and prepare base of dam

Centre line of dam

Limit of foundation

Stream bed

Limit of foundation

Diversion ditch

Dig out at least 0.3 m below old stream channel and remove sand, gravel, sediment and organic material

Barrage to prevent backflow

Centre line of dam

Anchor trench 1 m wide × 0.4 m deep

Impermeable

Centre line of dam

Cut-off trench 1.5 m wide extending 0.3 m into impermeable layer

Permeable layer

Impermeable layer

13. Build the water outlet structure(s), as necessary (see **Pond construction, 20/2**). Preferably, place it (or them) out of the stream bed, at a point to be dug lower than the lowest point in the pond.

Note: if the dam is to be constructed by machine, for example a bulldozer, the outlet structure could be built later (see pages 301 to 303).

14. Clearly **mark the construction height** of the dam and the crest width (refer to centre line) with stakes and lines, on the basis of planned dam characteristics (see Section 61). Maximum height is at the lowest point of the valley. Check the limits of the future pond upstream (see Section 112, **Topography, 16/2**).

15. **Set out the dam earthwork using** templates at intervals of 25 m or less and clearly showing the slopes of the sides. You can also use strings. If you use machinery, it is best to establish an auxiliary base line outside the radius of operation of the machinery, based on topographical survey bench-marks.

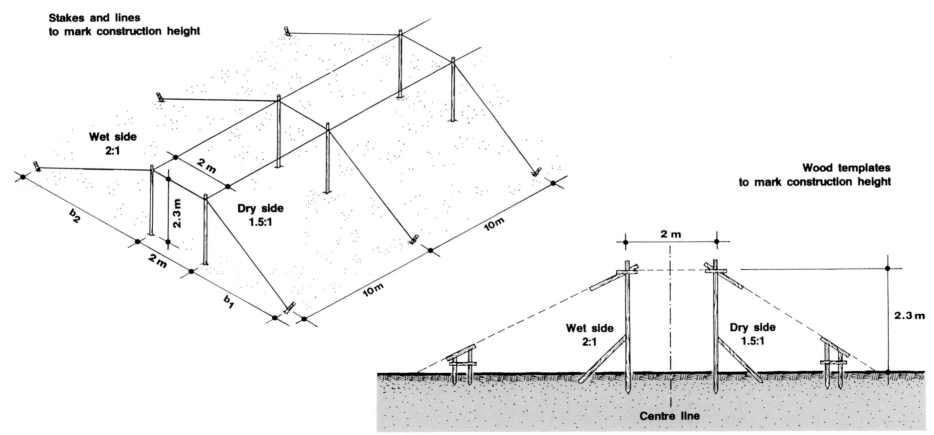

Stakes and lines to mark construction height

Wet side 2:1

Dry side 1.5:1

b_2

b_1

2 m

2 m

2.3 m

10 m

10 m

Wood templates to mark construction height

2 m

Wet side 2:1

Dry side 1.5:1

2.3 m

Centre line

16. Start building the dam by **laying out successive horizontal layers 15 to 25 cm thick**. Work across the entire site, from one side of the valley to the new stream channel, and from the wet to the dry side of the dam. Wet the soil if necessary and compact each layer well (see Section 62).

Build up dam using horizontal layers

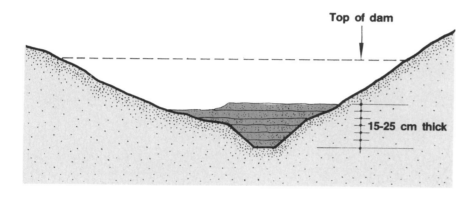

Top of dam

15-25 cm thick

Build each layer of good quality soil

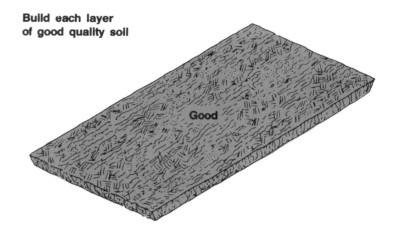

Good

If you do not have enough good soil to build the entire dam, use what good soil you have to build a central core

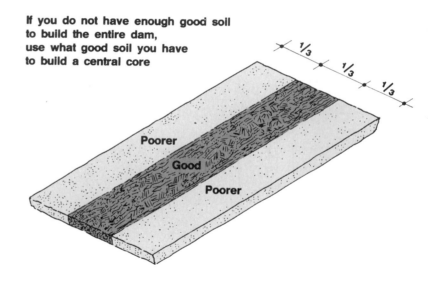

1/3 1/3 1/3

Poorer

Good

Poorer

17. According to the availability of clayey soil, you will use either **homogeneous soil layers** as wide as the dam or **heterogeneous soil layers**, each kind of soil covering only part of the dike width. Clearly mark the limits to be followed with stakes and lines.

(a) **If there is enough good soil** from which to build the entire dam, proceed by placing layers to cover the full base width.

(b) **If the supply of good soil is limited**, use it only to build a central core with the following characteristics:

- **width:** about one-third of the width of the dike;
- **side slopes:** at least a ratio of 1.5:1;
- **height:** water depth plus 20 cm.

Note: do not place heaps of soil next to each other without spreading them into a continuous layer before compaction.

Note: this core should be continuous with either the cut-off trench or the anchoring trench built in the foundations of the dam (see earlier) and should be properly placed and compacted.

(c) If you have to use various types of soil to build the dam, use the most impermeable material as a central core. Place the most permeable material on the dry side of the dam. Place the intermediate quality material on the wet side of the dam. Adjust each side slope to the particular kind of material used.

(d) If you do have relatively permeable materials on the dry side of the dam, it is useful to place larger grades (such as medium to coarse gravel or small rocks) at the dry toeline. This acts as a filter and prevents seepage water from washing out the finer dike material.

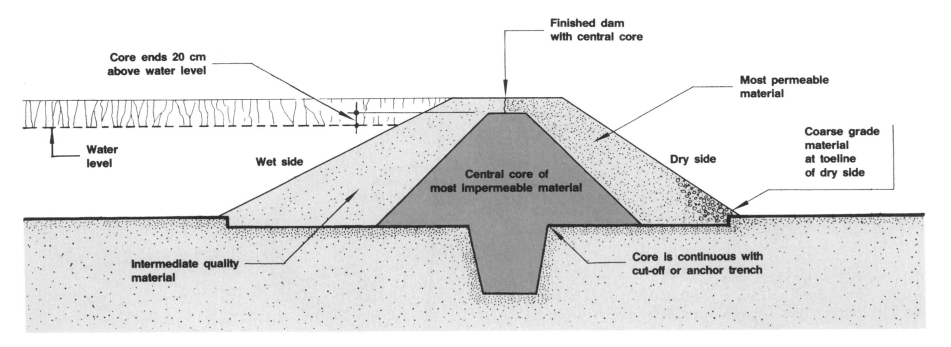

Finished dam with central core

Core ends 20 cm above water level

Most permeable material

Coarse grade material at toeline of dry side

Water level

Wet side

Dry side

Central core of most impermeable material

Intermediate quality material

Core is continuous with cut-off or anchor trench

Beware: you should pay particular attention to the compaction of the soil placed around the water outlet structures. Use good soil, the right moisture content, thin soil layers and thorough, strong tamping.

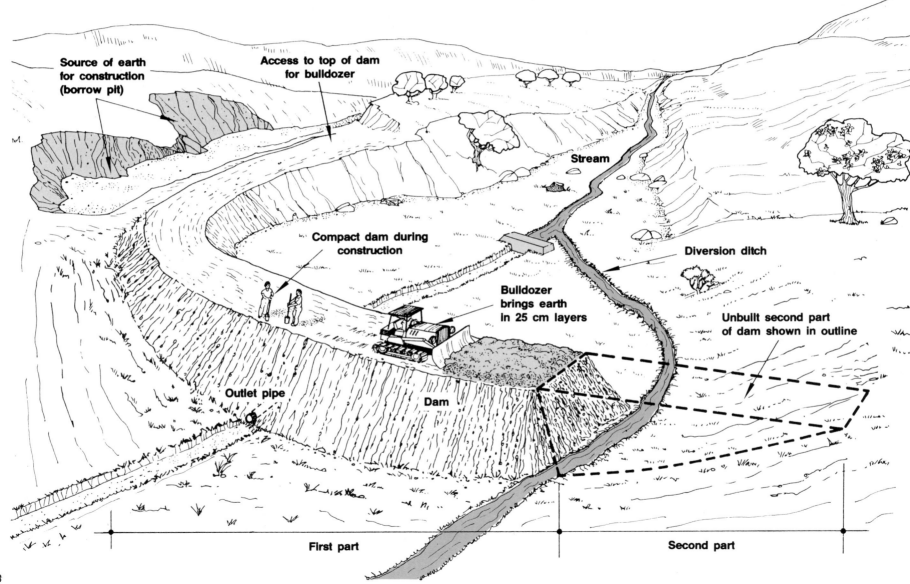

A typical barrage pond
built in two parts

Source of earth
for construction
(borrow pit)

Access to top of dam
for bulldozer

Stream

Compact dam during
construction

Diversion ditch

Bulldozer
brings earth
in 25 cm layers

Unbuilt second part
of dam shown in outline

Outlet pipe

Dam

First part

Second part

298

18. To build up the dam, you can use hand labour or machinery. **If you use machinery** such as a bulldozer for pushing, spreading and compacting the soil material, you could proceed in the following way.

(a) **Build up the first part of the dam** to about 1 m above the level of the foundations, progressing layer by layer.

(b) **Determine and stake out the centre line of the water outlet structure**, perpendicular to the centre line of the dam.

(c) **Mark a parallel line** on each side of this outlet centre line at a distance of about 0.5 m.

(d) **Dig a trench** about 1 m wide, down to the planned elevation for the water outlet pipes.

**Building a dam
for a barrage pond**

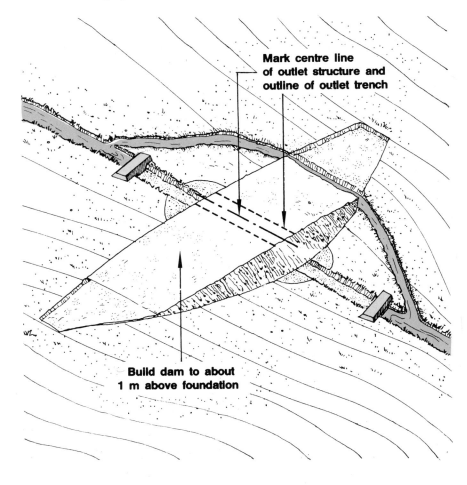

Mark centre line
of outlet structure and
outline of outlet trench

Build dam to about
1 m above foundation

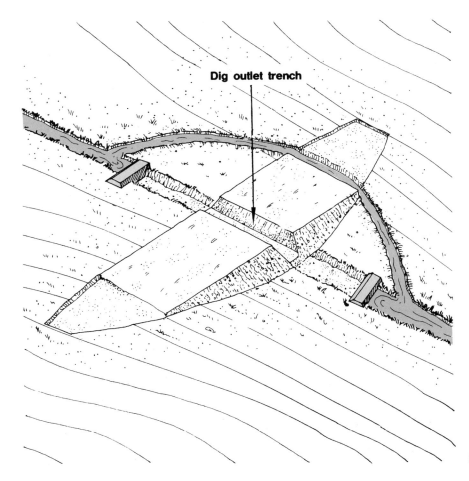

Dig outlet trench

Building a water outlet structure in the outlet trench

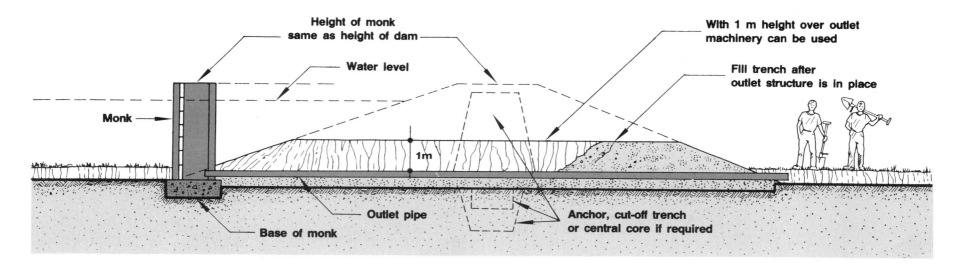

Height of monk
same as height of dam

Water level

With 1 m height over outlet
machinery can be used

Monk

Fill trench after
outlet structure is in place

1m

Outlet pipe

Base of monk

Anchor, cut-off trench
or central core if required

(e) **Build the water outlet structure** (see **Pond construction, 20/2**), being careful to make proper reinforcements around the areas where the water flows in and out of the structure.

(f) **Refill the trench** and **compact it well**, rebuilding the dam section as it was before. Pay particular attention to the central core if you have built one. Check carefully the quality of compaction around the water pipes. If possible use antiseepage collars (see **Pond construction, 20/2**).

(g) **Proceed with the building of the dam as before**. The water pipes are now well protected by 1 m of earth, and they will not collapse under the weight of the bulldozer.

19. When you reach the planned contruction height for the dam, carefully begin to **form the two side slopes**. Use slope gauges to assist you in cutting each slope at its planned angle (see Section 61, paragraph 12).

Note: it may be necessary to build a spillway and an emergency spillway (see Sections 113 and 114, **Pond construction, 20/2**).

A slope gauge

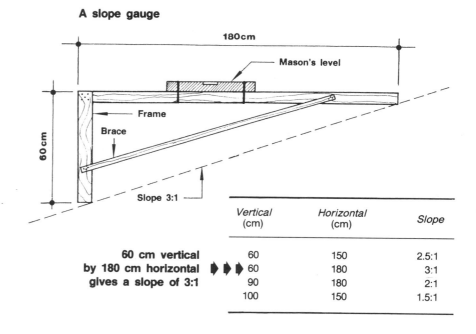

180cm

Mason's level

Frame

60 cm

Brace

Slope 3:1

	Vertical (cm)	Horizontal (cm)	Slope
60 cm vertical	60	150	2.5:1
by 180 cm horizontal ▶▶▶	60	180	3:1
gives a slope of 3:1	90	180	2:1
	100	150	1.5:1

20. **When the first part of the dam is completed**, let the stream run back into its old channel and through the water outlet structure. You are now ready to finish your barrage pond.

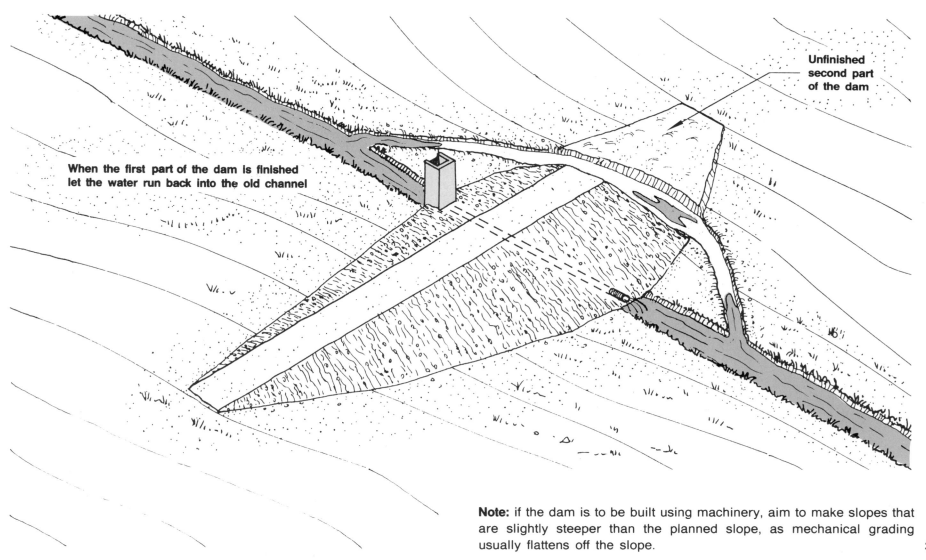

When the first part of the dam is finished let the water run back into the old channel

Unfinished second part of the dam

Note: if the dam is to be built using machinery, aim to make slopes that are slightly steeper than the planned slope, as mechanical grading usually flattens off the slope.

21. Repeat the previous operations for the second part of the dam in the area where the stream was diverted temporarily, but **first fill the bed of the diversion ditch** within the pond area.

(a) **Prepare good foundations**, extending them sideways well into the side of the valley. Be particularly careful to make a good foundation in the bed of the stream diversion.

(b) **Set out the dam earthwork** properly.

(c) **Build up the second part of the dam**, being particularly careful to connect it strongly to the first part of the dam and well into the side of the valley.

(d) **Form the two side slopes**.

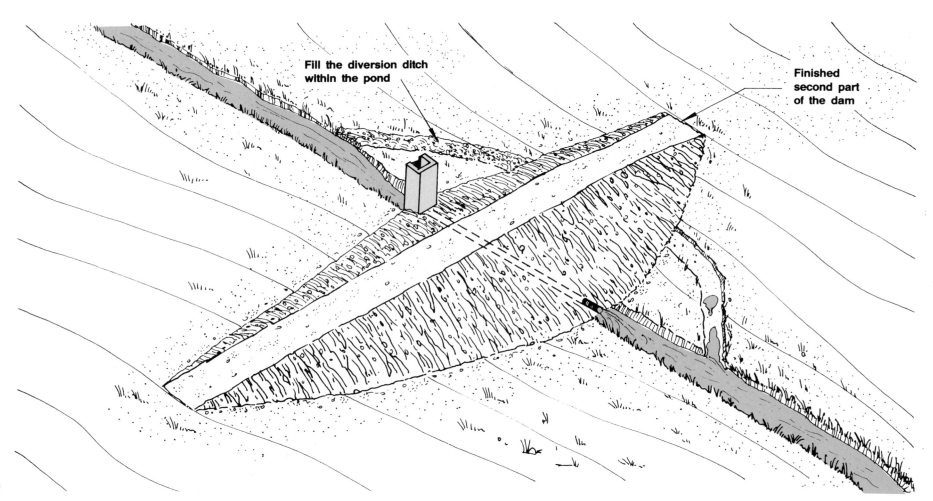

Fill the diversion ditch within the pond

Finished second part of the dam

22. **Work on the bottom of the pond** to make sure the pond is completely drainable.

(a) **Tidy up and shape** the course of the old stream bed.
(b) **Build a regular slope** toward the water outlet and dig bottom-drainers (see Section 610).

(c) If there are depressions, **dig a draining trench** toward a lower part of the pond bottom. This is important if you have borrowed the soil from within the pond area.
(d) If necessary, **fill any undrainable depressions**.

23. Finish your barrage pond by taking back some **surface soil**, spreading it on the dam and planting grass (see Section 69).

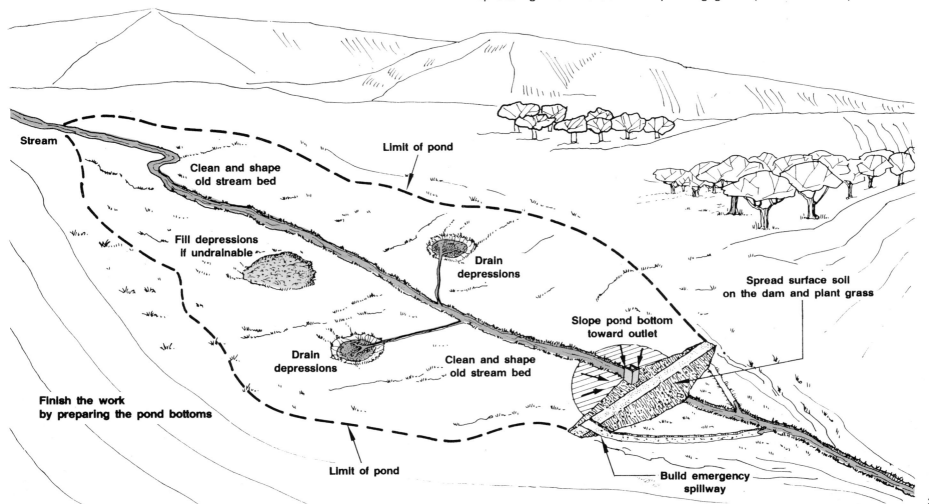

Stream

Clean and shape
old stream bed

Limit of pond

Fill depressions
if undrainable

Drain
depressions

Spread surface soil
on the dam and plant grass

Drain
depressions

Slope pond bottom
toward outlet

Clean and shape
old stream bed

Finish the work
by preparing the pond bottoms

Limit of pond

Build emergency
spillway

67 Constructing paddy ponds

1. Paddy ponds are embankment ponds built over flat ground. They have **four dikes of approximately equal height**. The size of the dikes, and thus the volume of earthwork, is usually limited to the minimum because of the need to import the soil material or to find it near the site.

Note: whenever the soil to build the dikes is taken by reducing the level across the whole pond floor, the resulting pond will be considered a cut-and-fill pond built on horizontal ground (see Section 68).

2. In some cases earth to build paddy ponds can be taken from areas next to the dikes, either inside or outside the pond, thus reducing construction costs. Trenches for dike material should not be cut with side slopes steeper than the dike itself.

How to remove soil material close to the dikes

Building the dikes using soil material from another place

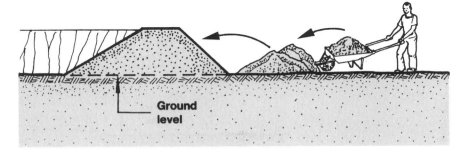

Ground level

Building the dikes using soil material from the site

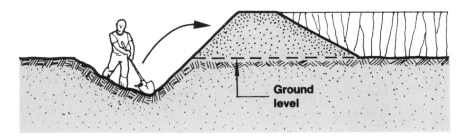

Ground level

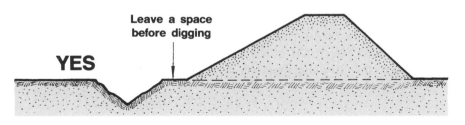

YES

Leave a space before digging

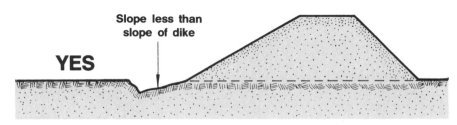

YES

Slope less than slope of dike

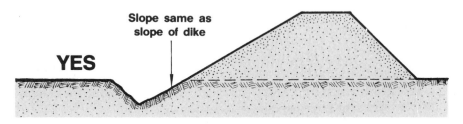

YES

Slope same as slope of dike

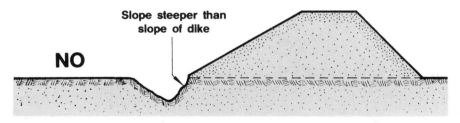

NO

Slope steeper than slope of dike

3. Clearly **mark the centre line** of each of the four dikes; the shape of the pond is usually either square (minimum earthwork) or rectangular, and the four centre lines will meet at right angles (see Section 36, **Topography**, **16/1**).

4. On each of the centre line stakes, indicate the level corresponding to the construction height **CH** of the dike to be built. Determine the level using one of the levelling methods described in **Topography**, **16/1**.

5. According to the characteristics of your dikes, calculate the width of each part of the dike base on either side of the centre line, as equal to:

$$\boxed{(\text{crest width} \div 2) + (\text{CH} \times \text{side slope})}$$

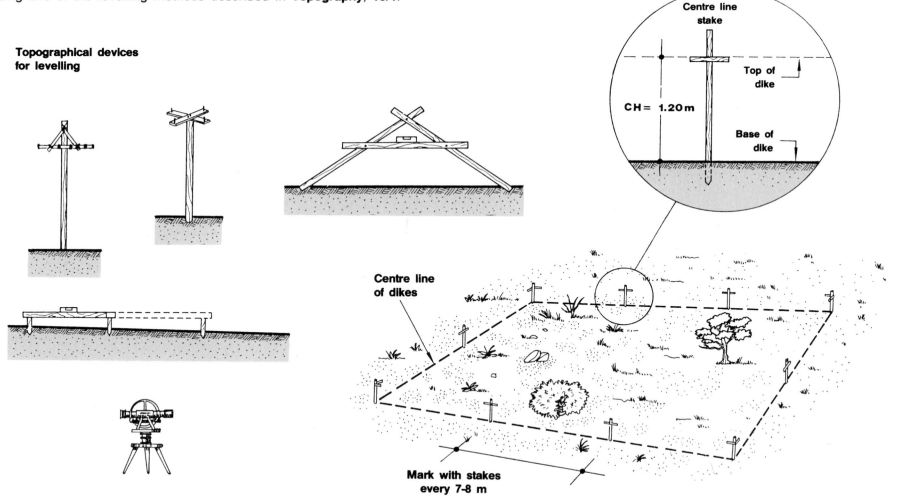

Topographical devices for levelling

Centre line stake

Top of dike

CH = 1.20 m

Base of dike

Centre line of dikes

Mark with stakes every 7-8 m

Example

If the crest width of dikes = 1 m, construction height = 1.20 m, wet side slope = 2:1 and dry side slope = 1.5:1, then at corner A distances AG and AH = (1 m ÷ 2) + (1.2 m × 2) = 0.5 m + 2.40 m = 2.90 m; distances AE and AF = (1 m ÷ 2) + (1.2 m × 1.5) = 0.5 m + 1.80 m = 2.30 m; and similarly all around the pond area for the other corners B, C and D.

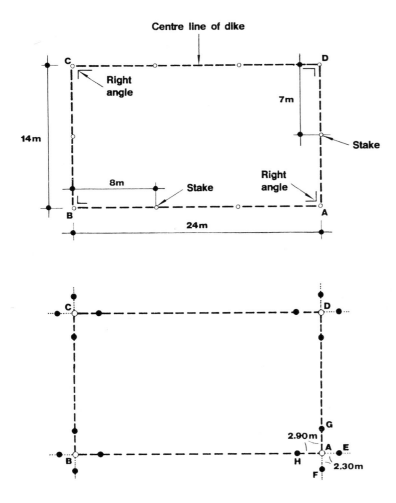

6. **Stake out the inside and outside limits of the dike base** by measuring these distances on perpendicular lines along the centre lines of the dikes and by setting out straight lines between these new points (see Section 16, **Topography**, **16/1**). In this way the base of each dike is clearly marked on the ground.

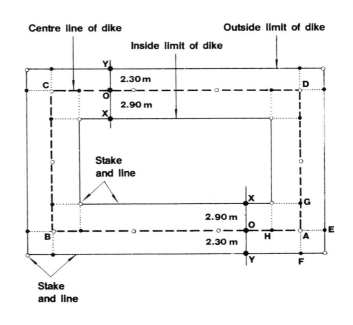

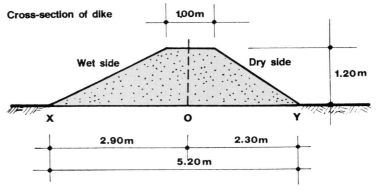

307

7. When the inner and outer limits of the pond have been staked, **clear any remaining vegetation** from the area (see Sections 52 to 55).

8. **Remove the surface soil** only from the area of the dike bases, as staked out above, and store it close by (see Section 56).

9. **Treat the surface of the foundations** of the dikes (see Section 63).

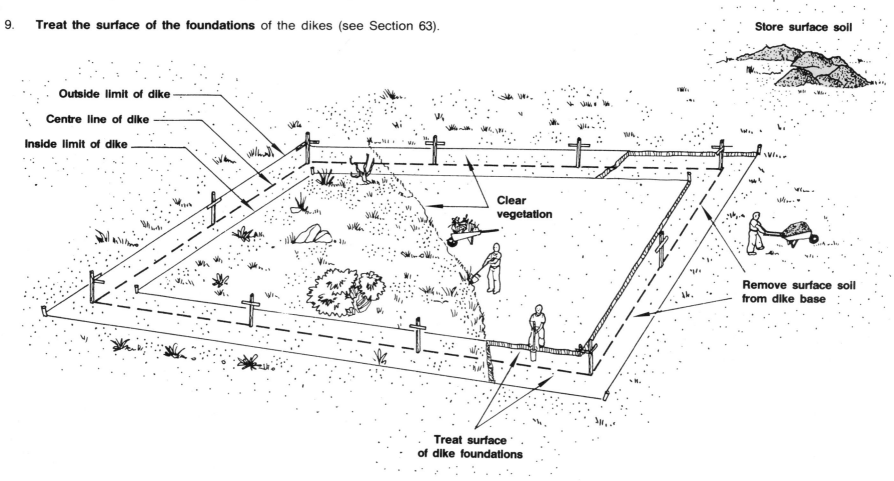

Store surface soil

Outside limit of dike

Centre line of dike

Inside limit of dike

Clear vegetation

Remove surface soil from dike base

Treat surface of dike foundations

10. According to local soil quality, **build either an anchoring trench or a cut-off trench** (see Section 63) along the centre lines of the dikes.

11. **Build the water control structures**, as necessary (see **Pond construction, 20/2**). Place the outlet entrance at an elevation low enough to ensure complete drainage of the pond along the sloping bottom (see paragraph 14 of this section).

Pipe inlet

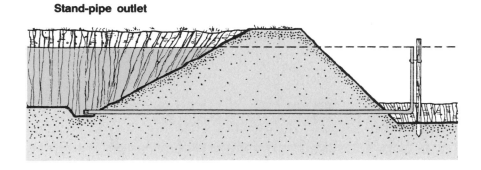

Stand-pipe outlet

Earthen canal inlet

Monk outlet

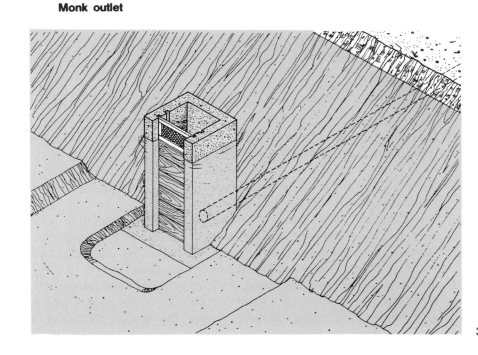

12.   There are several ways to build the dikes of a paddy pond. To construct them manually, you can use templates as for the barrage pond, although here only one template size is required.

**Wood templates
to mark construction height**

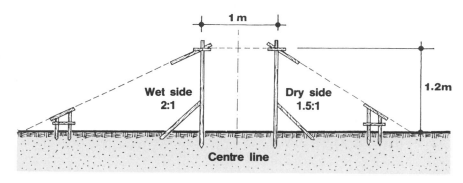

Note: you can also use stakes
and lines to mark construction height
as shown on page 295

13.   Another way to build the dikes of a paddy pond is given here.

(a)   Lay out a line to join and clearly mark the stakes setting out the inside limits of the dike's base. Attach this line at a height of about 0.20 m above the level of the surface of the dike's foundations.

(b)   Similarly lay out a line at the same level, joining the stakes setting out the outside limits of the dike's base.

(c)   Build the first layer of the four dikes 0.20 m high, bringing in good soil, placing it between the two strings all around the pond area, spreading it well, wetting and mixing it if necessary and tamping it thoroughly, especially around the outlet structure.

(d)   For each dike, move the inside limits of the dike base (stakes and lines) toward the centre line of the dike by a distance equal to 0.20 m × side slope; similarly, move the outside limits by a distance equal to 0.20 m × dry slope.

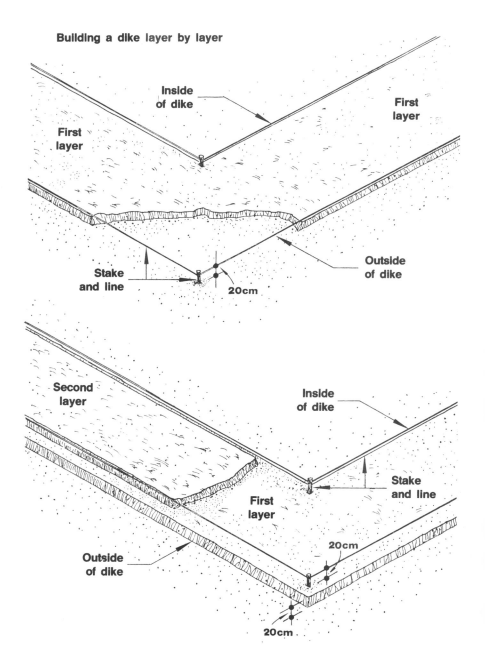

Building a dike layer by layer

If wet slope is 2:1 and dry slope is 1.5:1, move the inside limit by 0.20 m x 2 = 0.40 m and the outside limit by 0.20 m x 1.5 = 0.30 m.

(e) Raise all the lines 0.20 m higher.

(f) Build the second layer of the four dikes 0.20 m high between these new limits, as you did for the first layer.

(g) For each dike, move the inside and outside limits toward the centre line by the same distances as previously.

(h) Raise all the lines 0.20 m higher again.

(i) Build the next layer of the four dikes 0.20 m high between these new limits.

(j) Repeat these last three steps until you reach the level of the top of the dikes as indicated on the centre line stakes. It may be that the last soil layer is less than 0.20 m thick, in which case adjust the level of the lines with the level of the top of the dike.

Note: if you need to build a central core within the dikes, you should add lines setting out its width on either side of the centre line. The core is built together with the rest of the dike, using different types of soil for each 0.20-m layer.

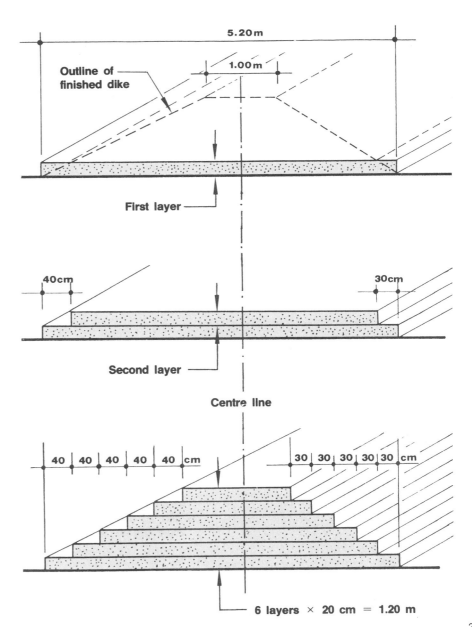

Finishing the dikes

14.   Now the dikes are built with sides that look like staircases. To reform these dikes with smooth side slopes and to finish their construction, proceed in the following way.

(a)   On the top of each dike, set out the planned dike crest width, measuring half of its value on either side of the centre line and marking the limits with wooden pegs and lines.

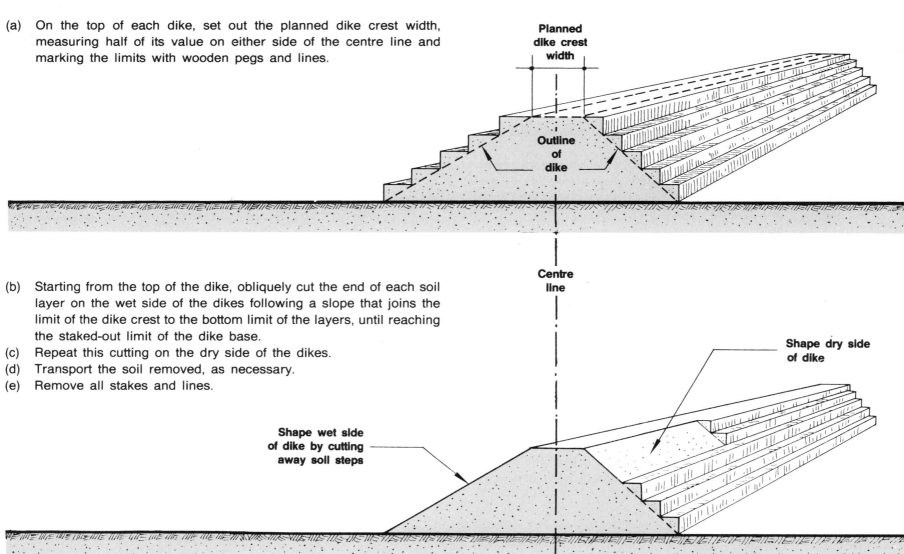

Planned dike crest width

Outline of dike

Centre line

(b)   Starting from the top of the dike, obliquely cut the end of each soil layer on the wet side of the dikes following a slope that joins the limit of the dike crest to the bottom limit of the layers, until reaching the staked-out limit of the dike base.
(c)   Repeat this cutting on the dry side of the dikes.
(d)   Transport the soil removed, as necessary.
(e)   Remove all stakes and lines.

Shape dry side of dike

Shape wet side of dike by cutting away soil steps

Cover top
and dry sides of dike
with surface soil

Surface soil

(f) Bring back some of the surface soil on the top of the dikes and on the dry sides.

Seed or plant grass
on top of dike
and on both sides
to the water's edge

Grass

Seed

(g) Seed or plant grass to control erosion (see Section 69).

15. The bottom of the pond now has to be finished, which is done using a levelling survey (see Chapter 5, **Topography, 16/1**).

16. For **smaller ponds**, give the bottom of the pond a gentle slope (0.5 to 1 percent) from the water inlet to the water outlet to ensure easy and complete drainage of the pond.

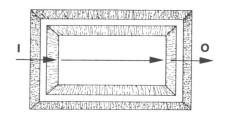

Bottom of pond (slope 0.5-1%)

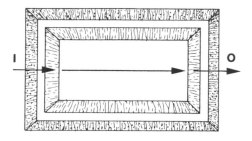

I = Inlet   O = Outlet

Note: you should always ensure that the entrance of the water outlet structure is at an elevation slightly lower than the lowest point of the bottom of the pond.

17. For **larger ponds** it is best to ensure complete drainage through a network of shallow drains each with a slope of 0.2 percent (see Section 610), rather than trying to build a slope over the entire pond area.

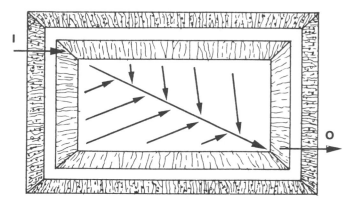

Bottom drains (slope 0.2%)

18. For ponds where internal trenches have been dug to provide material for dikes, these trenches should be connected together and shaped so as to drain toward the water outlet.

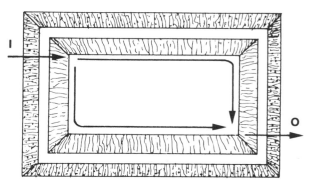

Internal trenches (slope 0.2%)

Building the dikes of a paddy pond using machinery

19.   When using machinery to build the dikes, a method similar to the one for barrage ponds can be used (see Section 65), except that four dikes are progressively built up instead of only one.

20.   It is best to establish an auxiliary base line with temporary bench-marks from which to set out the earthwork. This base line should be established outside the operation radius of the machinery.

21.   If the water outlet structure is built first, all pipes should be protected by a layer of earth at least 0.60 m thick to keep it from collapsing under the weight of the machinery.

Note: if a central core, an anchor or a cut-off trench is required, adapt the size(s) to the size of the dike.

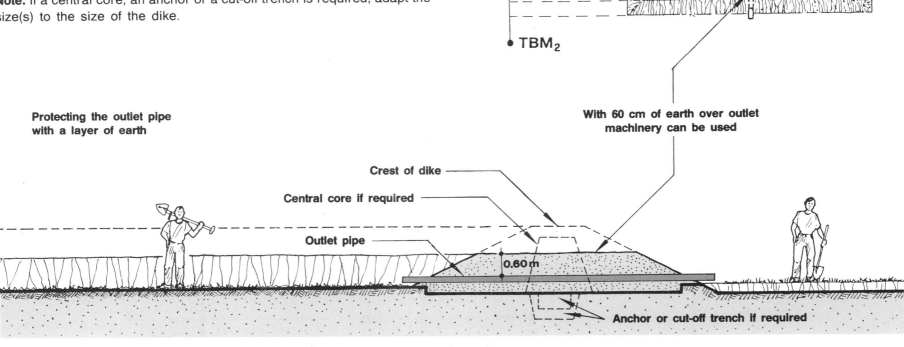

Protecting the outlet pipe
with a layer of earth

With 60 cm of earth over outlet
machinery can be used

Crest of dike

Central core if required

Outlet pipe

0.60 m

Anchor or cut-off trench if required

315

Building a series of adjoining paddy ponds

22. When building a series of adjoining paddy ponds, remember that only the **dikes forming the perimeter** of the pond series should be built with the characteristics of wet/dry dikes. The **intermediate dikes**, which are wet on both sides, can usually be built less strongly, even without a central core if necessary.

23. First stake out the centre line of a series of perimeter dikes, for example XY for dikes ACDB. Then stake out the centre lines of the opposite perimeter dikes EGHF and the perimeter or intermediate dikes AE, CG, DH and BF (see end of Section 37, **Topography**, **16/1**).

24. Build the dikes as explained earlier in this section, either manually (see paragraphs 13 and 14) or mechanically (see paragraphs 19 to 21).

Adjoining paddy ponds

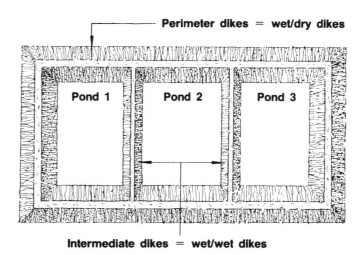

Perimeter dikes = wet/dry dikes

Pond 1   Pond 2   Pond 3

Intermediate dikes = wet/wet dikes

Stake out centre lines of dikes

316

Centre lines of
perimeter dikes

Pond 1   Pond 2   Pond 3

Centre
lines
of
intermediate
dikes

1.   In cut-and-fill ponds, at least some part of the pond dikes is made up by the natural ground, cut according to the planned side slope. Normally, a certain volume of soil necessary to build the dikes above ground level is obtained by digging a similar earth volume from inside the pond area, so as to obtain a pond with the planned depth (see cross-sections showing examples of earthwork in cut-and-fill ponds on page 318). The height of the dikes to be built is no longer equal to the depth of the pond, as in paddy ponds.

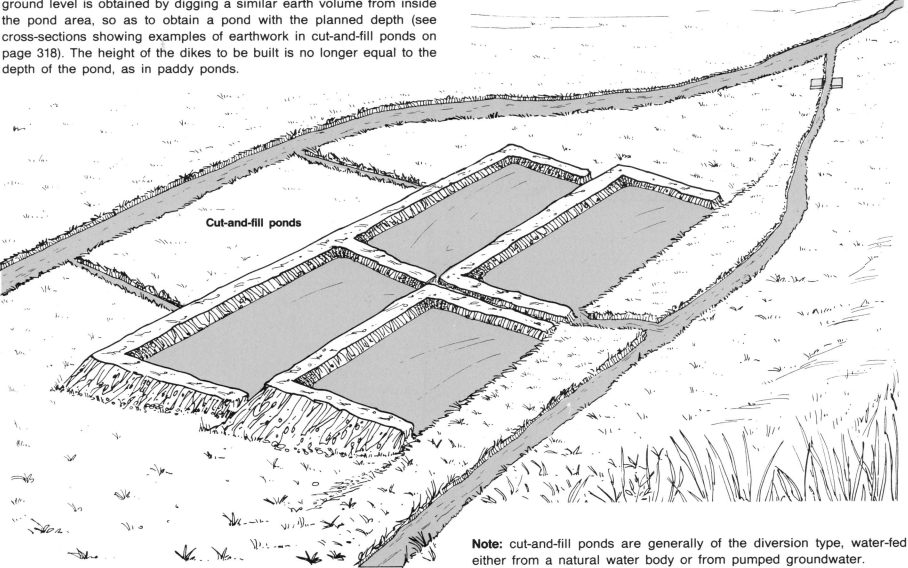

Cut-and-fill ponds

Note: cut-and-fill ponds are generally of the diversion type, water-fed either from a natural water body or from pumped groundwater.

Examples of earthwork in cut-and-fill ponds

Three-dike pond

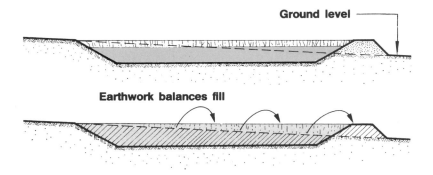

Ground level

Earthwork balances fill

Four-dike pond

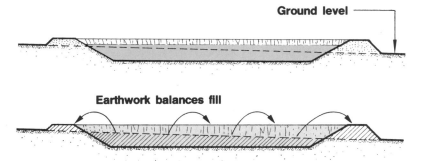

Ground level

Earthwork balances fill

Balancing cut-and-fill on horizontal ground

2.   During the planning stage, you must calculate how deep the pond should be dug to have enough soil material to build the four dikes and to create a pond with the planned depth. In good soils, this is usually done by matching the quantity of earth dug to the quantity needed to build the dikes. This is called **balancing cut-and-fill**. On horizontal ground, **two methods can be used for determining the balance of cut-and-fill.**

3.   In **method one**, excavation (cut) and dike (fill) volumes are calculated and approximately balanced by **trial and error using a graph**, as shown in the examples on pages 319 to 321. In **method two**, the balancing cut is determined using **a graph and tables**. Corresponding balancing volumes are then calculated, as described on page 322 and shown in the example on page 323.

4.   In practice there is no need for great accuracy, as it is not practical to control depths, heights and slopes with great precision. There will also be additional small volumes to allow for in making pond shapes and allowing space for inlets and outlets, access points, etc.

5.   To illustrate **method one** we will use a 400-m² pond (20 × 20 m along the centre lines of the dikes) with 2:1 inner slopes and 1.5:1 outer slopes, a dike construction height of 1.5 m and a crest width of 1 m.

Pond area = 20 × 20 m
        = 400 m²

Crest width = 1 m
      SW = 2:1
      SD = 1.5:1

Dike height = 1.5 m

Soil is taken from centre of pond to build dikes

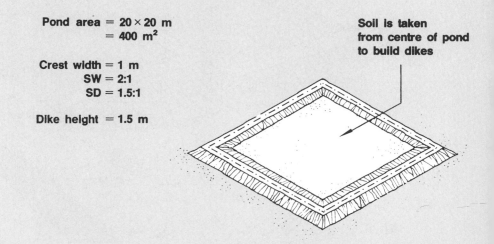

METHOD 1
First trial calculations

Example

If digging depth = 1 m, dike height = 1.5 m − 1 m = 0.5 m. Using the methods described in Section 64 obtain:

- cross-section of dike = (1 m × 0.5 m) + [(1 m × 0.5 m) ÷ 2] + [(0.75 m × 0.5 m)] ÷ 2] = 0.50 m² + 0.25 m² + 0.1875 m² = 0.9375 m²;
- total dike volume = 0.9375 m² × 80 m = 75 m³;
- average area of cut = [(169 m² + 289 m²) ÷ 2] = 229 m²;
- excavation volume = 229 m² × 1 m = 229 m³.

In this case, the cut volume greatly exceeds the fill volume.

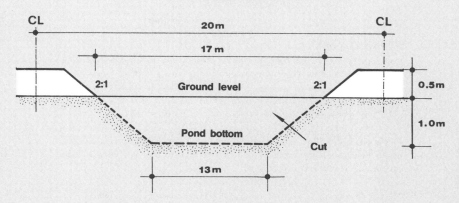

Pond cross-section

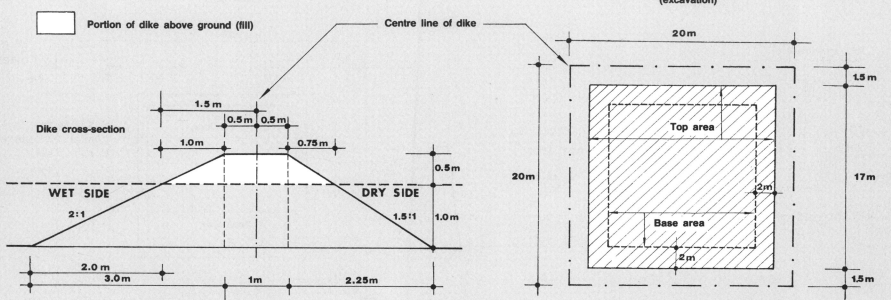

Plan of pond (excavation)

Dike cross-section

Portion of dike above ground (fill)

Centre line of dike

Top area of cut (ground level) = 17 × 17 m = 289 m²
Base area of cut (pond bottom) = 13 × 13 m = 169 m²

319

METHOD 1
Second trial calculations

Pond cross-section

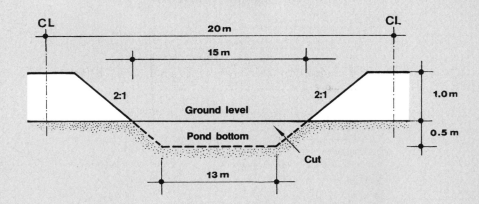

Example

Use higher dikes and reduce digging depth, say to 0.5 m; thus dike height = 1 m. This time, you obtain:

- total dike volume = 2.75 m² × 80 m = 220 m³;
- excavation volume = 197 m² × 0.5 m = 98.5 m³.

In this case, the fill volume exceeds the cut volume. The correct digging depth lies somewhere between 1 m and 0.5 m, the two values tried above.

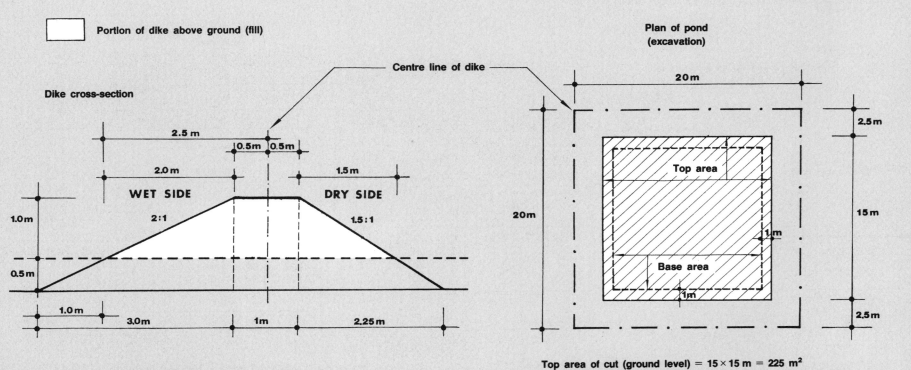

Top area of cut (ground level) = 15 × 15 m = 225 m²
Base area of cut (pond bottom) = 13 × 13 m = 169 m²

320

METHOD 1
Estimating correct digging depth

GRAPH 4

Two-way graph for cut-and-fill

Example

To estimate the correct digging depth, use a simple **two-way graph** (see **Graph 4**). Plot the excavation (cut) and dike (fill) volumes for trial 1 (points A and B) and trial 2 (D and C) respectively. Join AD and BC. The **intersection E** gives the **digging depth** required (0.72 m) and the approximate balancing volume (155 m³). You can check these results with a third set of calculations, where digging depth = 0.72 m and dike height = 1.50 m − 0.72 m = 0.78 m:

- total dike volume = 1.845 m² × 80 m = 147.6 m³;
- excavation volume = 210.6 m² × 0.72 m = 151.6 m³.

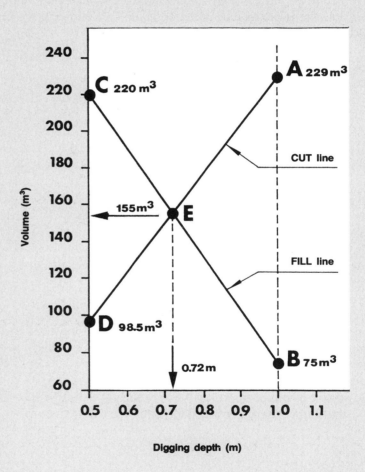

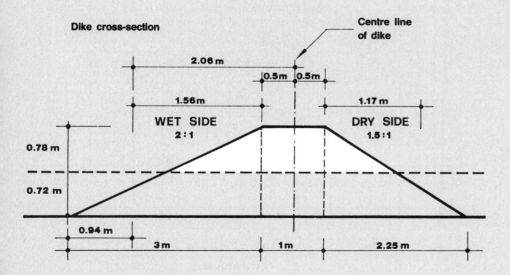

Dike cross-section

Centre line of dike

2.06 m

0.5m | 0.5m

1.56m | 1.17 m

WET SIDE 2:1 | DRY SIDE 1.5:1

0.78 m

0.72 m

0.94 m

3 m | 1m | 2.25 m

☐ **Portion of dike above ground (fill)**

6. Now let us look at **method two**, determined by using a graph and three reference tables.

7. This method is quick, but it is less accurate than method one. In addition, method two does not directly calculate the balancing volumes, although these can easily be calculated once the balancing cut is known. Proceed as follows.

METHOD 2

(a) In **Graph 5** enter the area of the pond (in m²). According to the width of the dike crest (in m) find the balancing cut depth (in m) for a **standard pond** where:

● the ratio of length to width is 1:1 (the pond is square);
● both dike slope ratios are 2:1;
● pond dikes are 1.5 m high.

(b) If the **side slopes** of your dike are not 2:1, correct the standard cut depth by **S** (in m), according to the first table given on page 323.
(c) If the **shape** of your pond is not square, multiply the cut depth by **P**, using the second table given on page 323.
(d) If the **height** of the dikes is not 1.5 m, multiply the cut depth by **D**, using the third table given on page 323.

GRAPH 5

Balancing cut depth for square ponds
(dike height = 1.5 m; inner and outer slope 2:1)

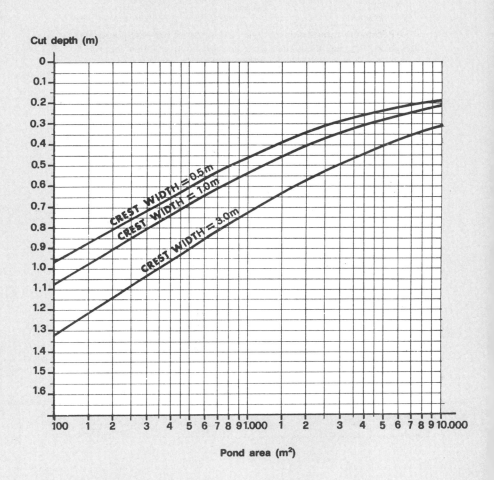

METHOD 2, continued

Example

Again using a pond 20 m square (area = 400 m²) with dikes 1.5 m high and a crest width of 1 m, from **Graph 5** determine a standard cut depth of 0.75 m.

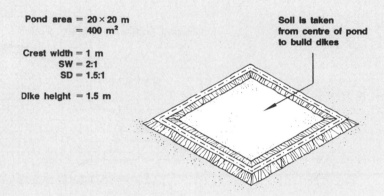

Pond area = 20 × 20 m
        = 400 m²

Crest width = 1 m
        SW = 2:1
        SD = 1.5:1

Dike height = 1.5 m

Soil is taken from centre of pond to build dikes

As the dike slope ratios differ from those of the standard pond (in this case inner slope 2:1 and outer slope 1.5:1) find, from the first table, **S** = −0.05 m and correct the standard cut depth as 0.75 m − 0.05 m = 0.70 m, which is the balancing cut.

If the pond had not been square, for example **L** = 28.5 m and **W** = 14 m, you should have found, from the second table, **P** = 1.04 and corrected the cut depth as 0.70 m × 1.04 = 0.728 m.

If the height of the dikes had been 2 m for example, you should have found, from the third table, **D** = 1.5 and corrected the cut depth further as 0.728 m × 1.5 = 1.092 m.

REFERENCE TABLES

1. Correction factor S for dike slope ratios

Inner slope	Outer slope	S (m)
1:1	1:1	−0.20
1.5:1	1:1	−0.15
1.5:1	1.5:1	−0.10
2:1	1.5:1	−0.05
2.5:1	1.5:1	−0.02
2:1	2:1	0
2.5:1	2:1	+0.03
3:1	2:1	+0.06
2.5:1	2.5:1	+0.08
3:1	2.5:1	+0.10

2. Correction factor P for pond shape

Pond length/width ratio	P
2	1.04
3	1.11
5	1.23
10	1.50

3. Correction factor D for dike height

Dike height (m)	D
1.0	0.55
1.2	0.74
1.4	0.9
1.5	1.0
1.6	1.1
1.8	1.3
2.0	1.5
2.2	1.8

Balancing cut-and-fill on sloping ground

8.   On regular sloping ground, the material needed for the dikes is also obtained from inside the pond area, but here both the height of the dikes above ground level and the digging depth vary according to the slope angle. This usually determines the position of the pond and hence the balancing depth.

9.   **If the ground slope is 0.5 percent at the most**, the site can be considered horizontal. If the pond is built with its length perpendicular to the contour lines and if dug to the same depth all round, the bottom of the pond will naturally have a 0.5 percent slope at the most.

10.   **If the ground slope is from 0.5 to 1.5 percent**, the pond should also be built with its length running across the contour lines, but the height above ground level of the two longer dikes will vary from one end to the other. Similarly, the width of these dike bases also varies. The **downslope dike will be the highest** above ground level, and the **upslope dike will be the lowest** above ground level. **Digging depth is the reverse**: it is greatest at the upslope end, least at the downslope end.

Digging ponds on sloping ground

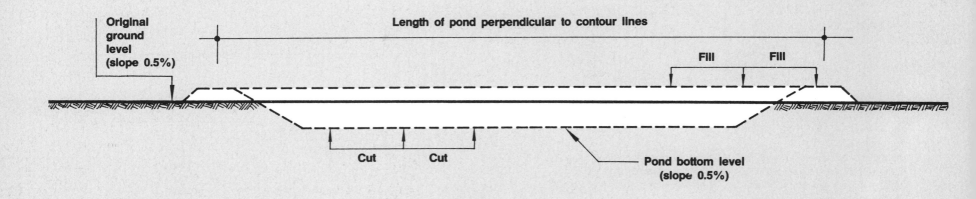

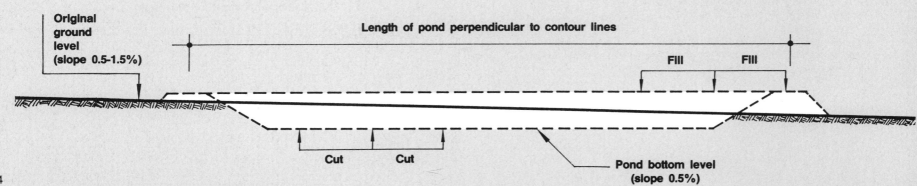

324

11. **If the ground slope is greater than 1.5 percent**, the pond should be built with its length running along the contour lines. The height above ground level of the two shorter dikes will vary from one end to the other. Similarly, the dike base width also varies. The longer dike downslope will be the highest above ground level. The longer dike upslope will be the lowest above ground level. Digging depth is the reverse: greatest in the upslope part of the pond, least in the downslope part.

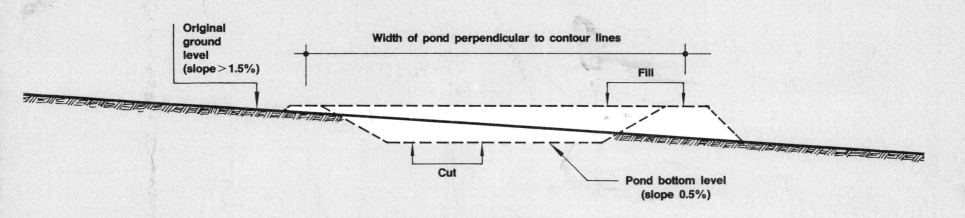

12. To obtain a rapid estimate with any slope values, you can use either of the previous two methods.

(a) **Method 1**, the trial-and-error method, uses the volume calculations for horizontal ground with average ground level and average dike height figures.

(b) **Method 2**, for horizontal ground, uses average ground level and average dike height figures.

Note: these methods are accurate enough when the slope is less than 0.5 percent.

13. To obtain a **better estimate** of the balancing depth of cut **in more steeply sloping ground** (more than 0.5 percent) you should use **Method 1** together with the methods for calculating excavation and dike volumes on sloping ground (see paragraphs 13 to 17, Section 64).

14. The details of this procedure will vary according to the ground slope.

15. On **gentle slopes** (0.5 to 1.5 percent), you will have different types of dike:

- **one low shorter dike**, upslope, either horizontal or varying in height;
- **one high shorter dike**, downslope, either horizontal or varying in height;
- **two longer dikes**, varying in height.

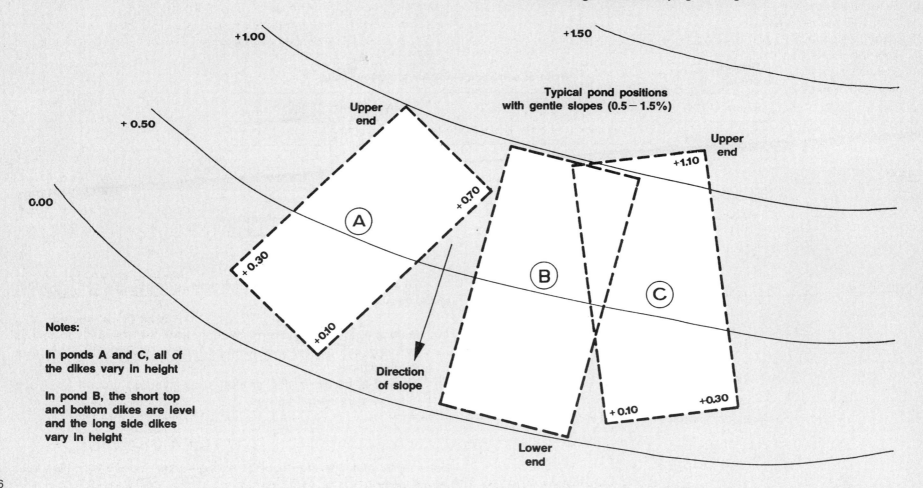

Typical pond positions with gentle slopes (0.5 – 1.5%)

+1.00

+1.50

Upper end

+ 0.50

Upper end

0.00

+0.70

+1.10

A

+0.30

B

C

+0.10

Direction of slope

+0.10

+0.30

Lower end

Notes:

In ponds A and C, all of the dikes vary in height

In pond B, the short top and bottom dikes are level and the long side dikes vary in height

16. **Apply Method 1** in the following way.

(a) **Select** a first minimum depth of cut measured at the lower end of the pond; calculate excavation volume using the method described in Section 64, paragraph 23 (see pages 279 and 280).

(b) **Calculate** the corresponding dike volume using the method described in Section 64, paragraph 14 (see page 276).

(c) **Plot** these values on the two-way graph (see **Graph 4**).

(d) **Select** a second minimum depth of cut and calculate excavation and dike volumes similarly.

(e) Again, **plot** these values on the two-way graph (see **Graph 4**).

(f) **Join** the points **A** to **D** and **C** to **B** and mark the intersection point **E**. This will determine the balancing minimum cut and the corresponding dike volumes.

17. For **ground slopes greater than 1.5 percent**, you will have:

● **one low longer dike**, upslope, either horizontal or varying in height;
● **one high longer dike**, downslope, either horizontal or varying in height;
● **two shorter dikes**, varying in height.

18. Select minimum depth of cut and calculate dike volumes accordingly. Complete the trial-and-error procedure using **Method 1** for sloping ground as just described (see page 318 and the examples following it on pages 319 to 321).

Typical pond positions with steep slopes (greater than 1.5%)

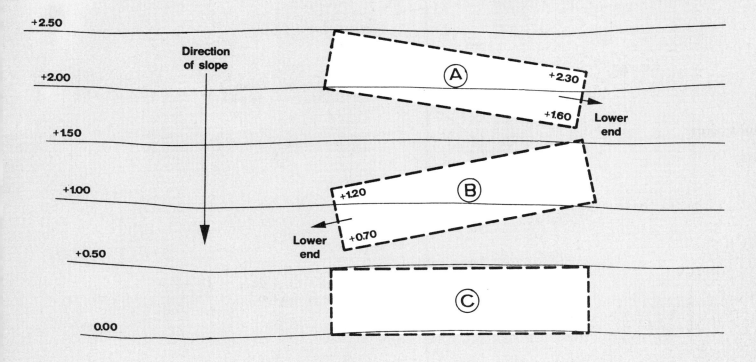

Notes:

In ponds A and B, all of the dikes vary in height

In pond C, the long top and bottom dikes are level and the short side dikes vary in height

The lower ends of ponds A and B face in opposite directions, because of their angled position on the contour

327

19. If the construction site of your pond is characterized by **irregular slopes and uneven ground**, the problem of balancing cut-and-fill earthwork volumes becomes much more complicated. According to the extent of the earthwork involved, rough estimates may be obtained by using **Method 1** with volumes calculated as described in Section 64.

● for small ponds it is best to use average ground level;

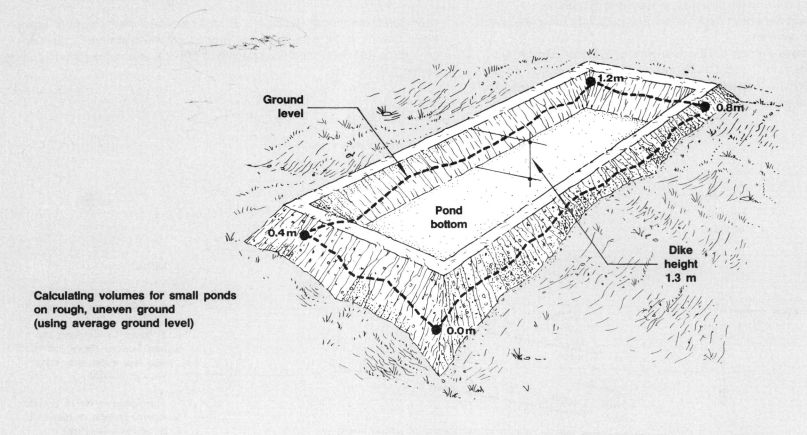

Calculating volumes for small ponds on rough, uneven ground (using average ground level)

Average ground level = (1.2 + 0.8 + 0.0 + 0.4 m) ÷ 4 = 0.6 m
Average dike height = 1.3 − 0.6 m = 0.7 m

- for larger ponds more accurate results are obtained by calculating volumes section by section.

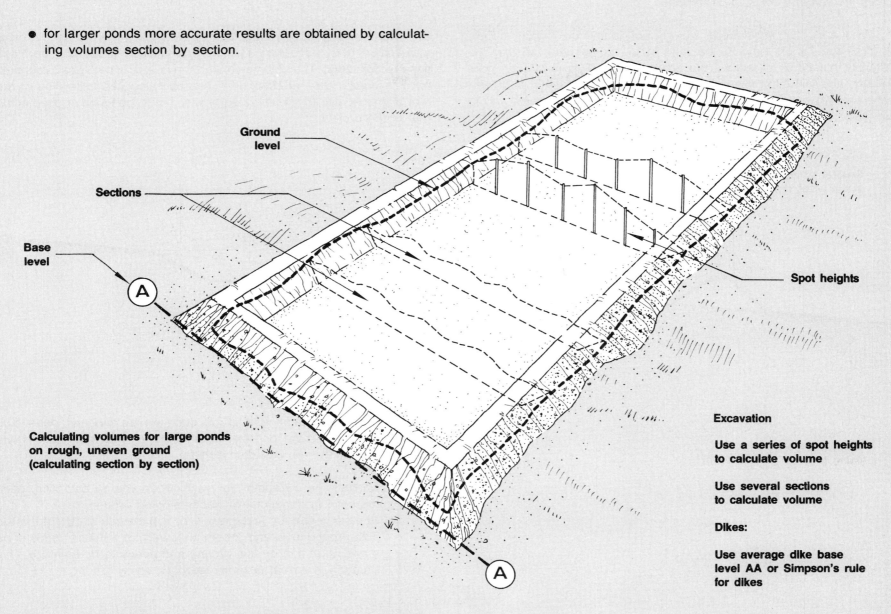

Ground level

Sections

Base level

(A)

Spot heights

Calculating volumes for large ponds on rough, uneven ground (calculating section by section)

(A)

Excavation

Use a series of spot heights to calculate volume

Use several sections to calculate volume

Dikes:

Use average dike base level AA or Simpson's rule for dikes

20. Alternatively, **Method 2** using average ground level values can be applied. This method is faster, but less accurate.

21. Frequently, for larger sites, several ponds and their water supply and drainage canals have to be built at the same time, so that the earthwork balances out across the entire project. This is clearly more complex, and will often require the assistance of a qualified engineer. There are, however, a few ways to estimate your requirements and to help guide your decisions, as you will learn below.

22. **Water supply** and **drainage canals** are usually more or less fixed at a specified level necessary for them to function properly (see **Pond construction**, **20/2**). The volumes of earth involved, either excavated, built up or a combination of cut-and-fill, can be calculated according to the size of the canals. Any deficit or surplus soil must be found or used in the rest of the project.

Several ponds built at the same time

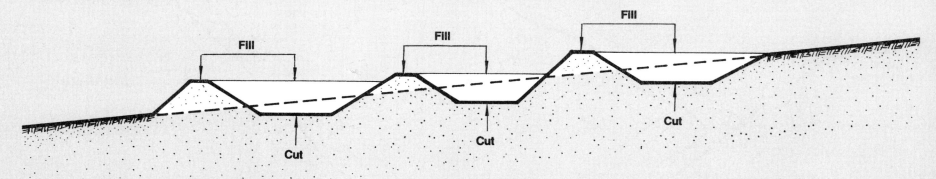

23. The site area can be split up into **main groups of ponds**, depending on location, pond type or size, or method of operation (see **Pond construction**, **20/2**). You can then decide:

● whether the cut-and-fill for each group will be balanced, often the case in horizontal or gently sloping ground; or
● whether a surplus is expected for a particular group (if it is on high ground level and should be lower), or a deficit needs to be made up (if it is on low ground and needs to be built up). This situation is typical in more steeply sloping ground.

24. For balancing groups of ponds, you could arrange for all of the **ponds to be at the same level**. This solution is appropriate for horizontal or gently sloping ground. Calculate the cut-and-fill balance using Method 1. Use average ground and dike levels for the group of ponds as a whole with the perimeter and intermediate dikes. Add in the extra cut or fill required for the canals.

Note: if the intermediate dikes are small, disregard them altogether and treat the whole group of ponds as a single large pond.

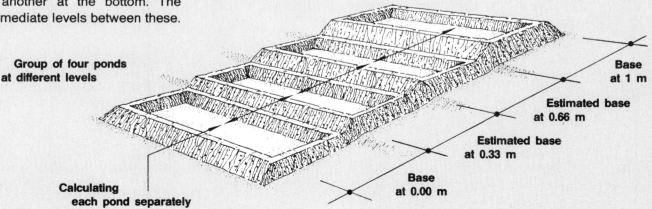

If ponds are calculated separately, allow for shared dikes

Group of six ponds at the same level

To simplify the layout, the group of six ponds can also be treated as a single large pond

Use average ground levels and dike heights for sloping ground

25. Alternatively you can arrange for **ponds to be at different levels**. This solution is appropriate for more steeply sloping ground. Here each pond is calculated individually. A simple short cut is to calculate one pond at the top end of the site and another at the bottom. The intermediate ponds will then be set at intermediate levels between these.

Group of four ponds at different levels

Base at 1 m

Estimated base at 0.66 m

Estimated base at 0.33 m

Base at 0.00 m

Calculating each pond separately

331

26. Where a surplus or a deficit is involved, you should include it in the trial-and-error balance:

- if a **surplus** is required, add it to the dike volume quantity. The total figure is then used in the two-way graph;
- if a **deficit** is to be made up, add the required amount of soil to be brought into the area to the excavation volume. This total figure is then used in the two-way graph.

27. In several areas certain groups of ponds may have **preset levels defined**, for example, by water supply or drainage. In this case, cut-and-fill calculations will define the surplus soil produced or the deficit needed to be made up from elsewhere.

28. Another useful method is to draw one or more cross-sections through the site or through groups of ponds. You can quite simply **adjust levels graphically** to obtain an approximate balance of cut-and-fill.

Approximate balance of cut-and-fill

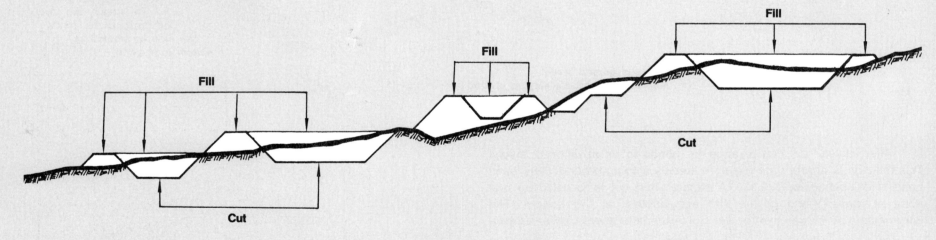

Staking out a cut-and-fill pond on horizontal ground

29. When the **ground slope is less than 0.5 percent**, the first steps of the method for staking out the centre lines of the dikes and the limits of the bases of the dikes are similar to the method described earlier for paddy ponds (see Section 67). The top of all the centre line stakes should be at the same level, indicating the dike's construction height.

30. In most cases, you will have made all the necessary calculations for staking out when calculating earth volumes.

Example

If crest width of dike = 1.00 m;
   pond depth = 1.50 m;
   dike height = 0.78 m;
   dry side slope = 1.5:1;
   wet side slope = 2:1.
Then, staking-out distances at ground level are:
   $Z = (1.00 \text{ m} \div 2) + (0.78 \text{ m} \times 1.5) = 1.67 \text{ m}$;
   $X = (1.00 \text{ m} \div 2) + (0.78 \text{ m} \times 2) = 2.06 \text{ m}$;
and similarly for the other three dikes.

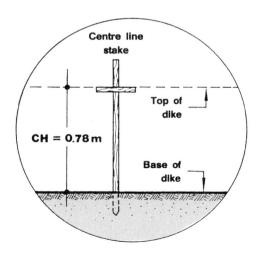

Centre line stake

Top of dike

CH = 0.78 m

Base of dike

Plan for staking dikes

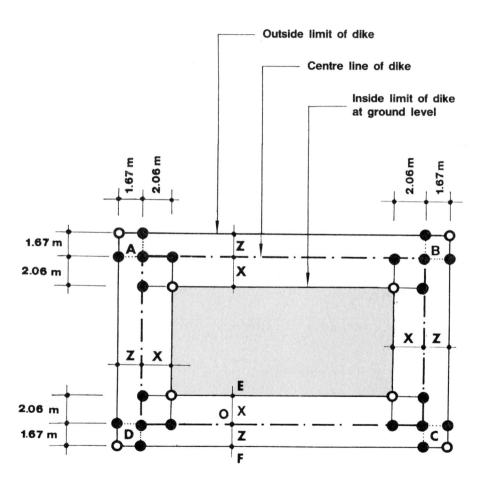

Outside limit of dike

Centre line of dike

Inside limit of dike at ground level

Dike cross-section

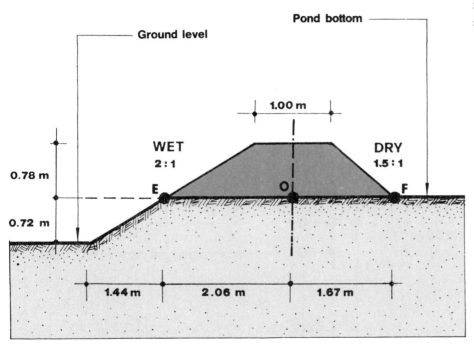

Portion of dike above ground (fill)

31. However, as the side slopes of the dikes have to be cut below ground level, you should stake out with short pegs one additional line defining the **limits of the pond bottom**.

Note: the pond bottom area is the same regardless of the cut depth (see Section 64).

32. Stake out the pond bottom to clearly indicate how deep to dig at a series of points (see Section 114, **Topography**, **16/2**).

Staking out pond bottom

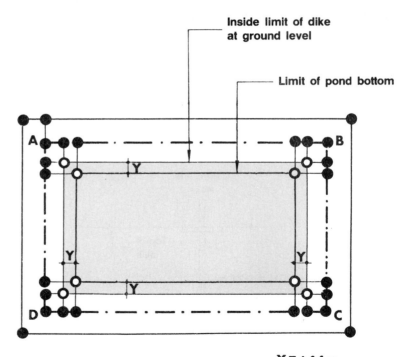

Y = 1.44 m

33. When the **ground slope is steeper than 0.5 percent**, you have already learned that the parts of the dikes to be built above ground level do not have the same height on every corner of the pond. The top of the dikes needs to be level, but as the base of the dikes is on a variable level, the width of the dikes at the base varies from one pond corner to another.

34. Once you have calculated how deep you need to dig at each pond corner to balance cut-and-fill volumes, the characteristics of the dikes are fully defined. In particular, their **height above ground** is determined for each pond corner, and therefore their corresponding base width is known. Now it remains to clearly mark these measurements on the ground before starting the construction. Proceed as shown in the example.

Example

The pond to be built measures 25 × 15 m along centre lines. The dike is 1.40 m high with the following characteristics:

- wet/dry side slopes = 2:1;
- top width (crest) = 1.00 m;
- the ground slope = 1.5 percent;
- the cut-and-fill volumes are found to balance for a minimum cut = 0.86 m;
- the maximum cut = 1.15 m;
- thus the maximum height of the dike to be built above ground level is 1.40 m − 0.86 m = 0.54 m, and the minimum height of the dike is 1.40 m − 1.15 m = 0.25 m.

This example is continued on pages 336 and 337.

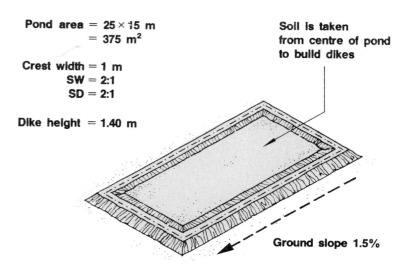

Pond area = 25 × 15 m
= 375 m²

Crest width = 1 m
SW = 2:1
SD = 2:1

Dike height = 1.40 m

Soil is taken from centre of pond to build dikes

Ground slope 1.5%

(a) Stake out the four centre lines of the dikes AB, BC, CD and DA at 25 × 15 m, the width of the pond being parallel to the contour lines.

Stake out centre lines ABCD

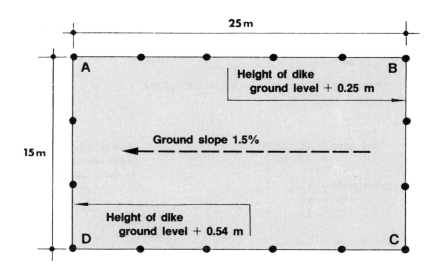

(b) Calculate for each pond corner the width of the dike base to be staked out on either side of its centre line as X or Z = (crest width ÷ 2) + (side slope × height above ground):

- **Corner A**
  inside distance X = (1 m ÷ 2) + (2 × 0.54 m) = 1.58 m;
  outside distance Z = (1 m ÷ 2) + (2 × 0.54 m) = 1.58 m;
- **Corner B**
  inside distance X = (1 m ÷ 2) + (2 × 0.25 m) = 1.00 m;
  outside distance Z = (1 m ÷ 2) + (2 × 0.25 m) = 1.00 m;
- **Corner C**
  as corner B = 1.00 m;
- **Corner D**
  as corner A = 1.58 m.

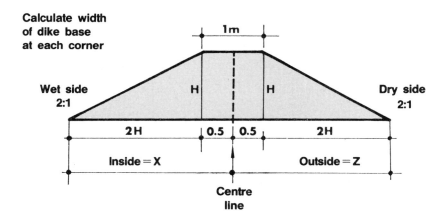

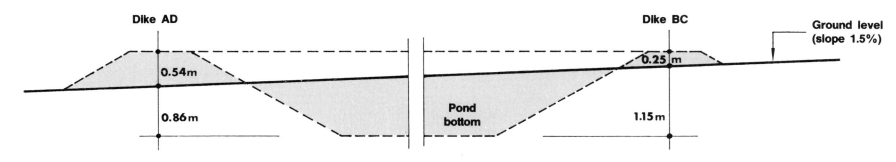

(c) Stake out these distances **X** and **Z** on either side of the centre lines at each pond corner and in two perpendicular directions to obtain four new points at each corner.

(d) Join these new points to set out the four **dike bases at ground level**. Notice that the limits of the side walls are not parallel, owing to the difference in elevation along the pond length.

(e) The pond bottom is set out in the same way as previously described.

Note: if the **ground slopes in more than one direction**, as when the pond is set at an angle across the slope, the same method can be used. But in this case, each of the pond corners is at a different height and therefore none of the pond walls are parallel.

Example of dike base on ground sloping in only one direction

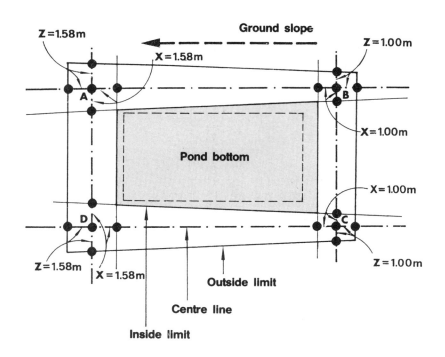

Example of dike base on ground sloping in more than one direction

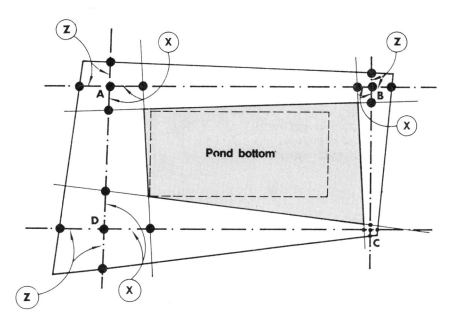

Staking out a cut-and-fill pond on a very irregular slope

35. If the **ground slope is very irregular**, it is best to proceed slightly differently.

(a) Stake out the centre lines with a series of pegs.

(b) On each of the pegs mark the height to be reached by the dike crest. This forms a horizontal line.

(c) Calculate the required dike base width at each peg according to the dike to be built there above ground level.

(d) Stake out the dikes' bases all around the pond, on short perpendiculars set on the centre lines.

(e) Stake out and mark with lines the limits of the pond bottom after calculating distances **Y** from the centre line as:

$$Y = (\text{crest width} \div 2) + [(\text{total dike height}) \times (\text{wet side slope})]$$

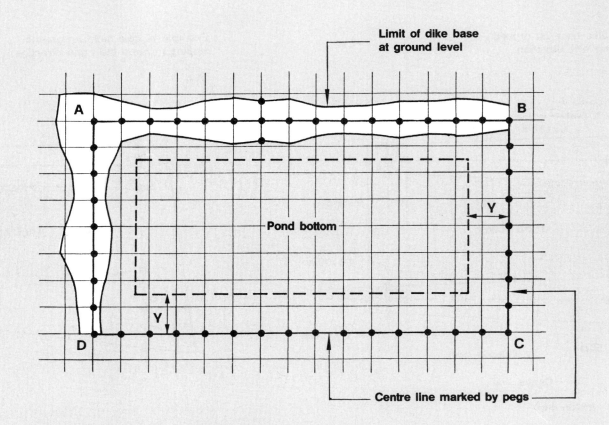

Limit of dike base at ground level

Example of dike base on very irregular ground

Pond bottom

Y

Y

A

B

D

C

Centre line marked by pegs

Building the dikes manually

36. Start digging within the area staked out as the pond bottom, cutting vertically along the edges of this area.

37. Throw this earth into the area staked out as the dike base. Spread it over the entire area into a layer about 0.20 m thick, wet it if necessary, and compact it well (see Section 63).

38. Build up and shape the dikes to ground level as described for paddy ponds (see Section 67), checking the pond bottom level from time to time.

39. Finish the dikes by cutting the earth left between the lines of stakes marking the inside limits of the base of the dikes at ground level and at pond bottom level. This also completes the wet side slopes.

40. Remove all stakes and lines, put back the surface soil on the dikes and plant or seed grass (see Section 69).

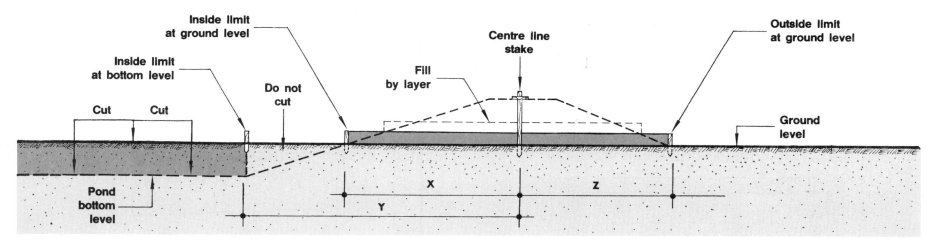

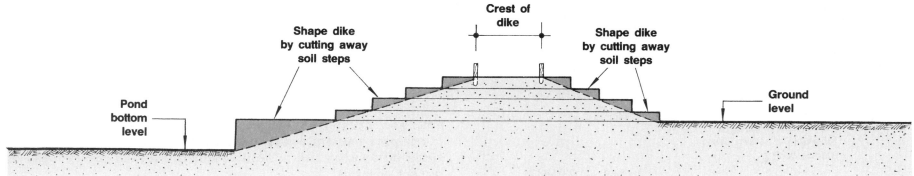

Completing the cut-and-fill pond

41. Clean the pond bottom.

42. In **small ponds**, give the bottom a gentle slope (0.5 to 1 percent) from water inlet to water outlet.

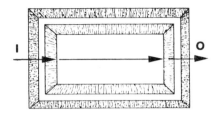

**Pond with sloping bottom
(slope 0.5 – 1%)**

43. In **large ponds**, either give the bottom a very gentle slope (0.2 percent) or, preferably, dig a network of shallow drains with a slope of 0.2 percent over the entire bottom area (see Section 610).

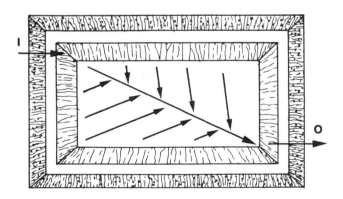

**Pond with a network
of shallow bottom drains
(slope 0.2%)**

I = Inlet   O = Outlet

Building the dikes using machinery

44. When using machinery to build the dikes of a cut-and-fill pond, it is most important to **check on the digging progress closely and regularly**, to avoid cutting the pond too deep. Usually the bulldozer transports the soil by pushing, spreading it into a thin layer over the dikes' area and compacting it. Compaction in particular should be thorough, after wetting if necessary.

**External reference markers
for use with machinery**

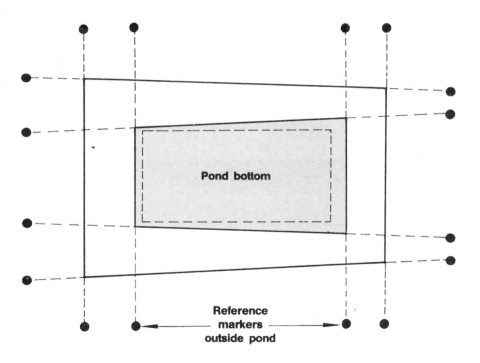

Pond bottom

**Reference
markers
outside pond**

Note: when machinery is being used, it is also useful to set up a number of **reference markers outside the pond** itself, as internal markers can easily be knocked over. These reference markers can be used to recheck positions.

340

69 Protecting dikes against erosion by rain

Protect new dikes as soon as they are built

1. Newly built dikes should be protected against erosion by **planting or seeding a grass cover** on the crest of the dikes, on their dry side and on their wet side down to the normal water level of the pond.

2. To form a grass cover with the minimum of delay, proceed as follows:

(a) Spread a 10- to 15-cm layer of topsoil over the area to be planted. This topsoil is either brought back to the pond area from which it was earlier removed or is obtained from a nearby source.

(b) If possible, mix in some compound chemical fertilizer, for example a 13-13-13 mixture (NPK)[1] at the rate of 50 to 100 g per m^2 surface area or 400-800 g per m^3 of topsoil.

(c) Plant either cuttings or pieces of turf of the selected grass, see **Table 30**, at relatively short intervals.

(d) Water well immediately after planting and afterwards at regular intervals.

(e) After the sod is formed, cut it short regularly to encourage it to spread all over the area. If possible, apply about 0.1 g of actual nitrogen per m^2 to accelerate the spreading.

3. For further advice, contact agricultural extension workers.

---

1) N = nitrogen; P = phosphorus; K = potassium

Preparing dikes for planting or seeding grass

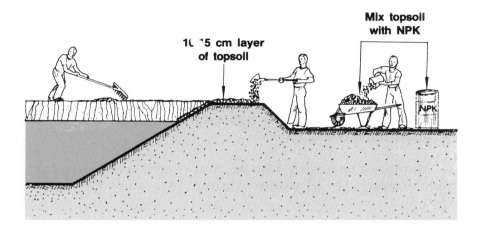

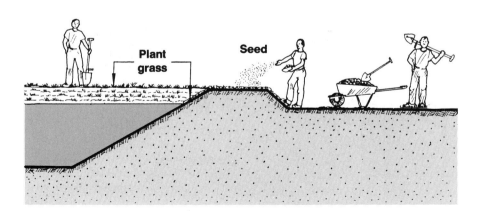

4. If the weather is dry, you should plan for **regular watering** of the newly planted grass. Use **mulching*** to reduce soil evaporation.

5. When it rains heavily, use **temporary protection**, such as hay or other suitable materials, to avoid severe erosion of the dikes until a grass cover is formed.

6. Never plant large trees on or near dikes because their roots would weaken the dikes. In some areas, **vegetable crops** and **forage bushes** can be grown, but care should be taken to select plants with a good ground cover and with roots that do not weaken the dike by penetrating too deeply or disturbing the soil.

7. Care should be taken to keep the dikes in good condition, and **only small animals** should be allowed to graze or browse on them.

Crops may be planted on top of dikes

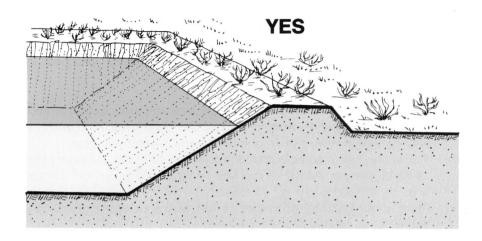

YES

Never plant large trees on or near dikes

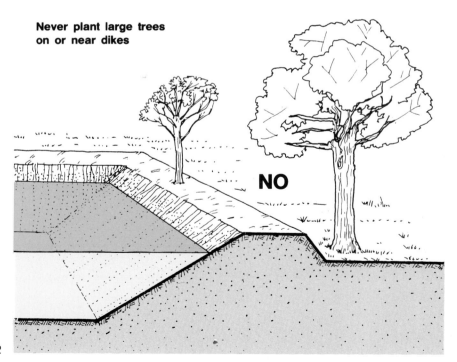

NO

Only small animals should graze on dikes

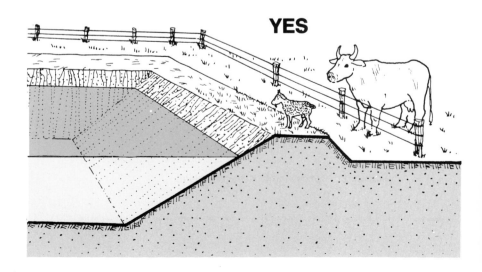

YES

Selecting the grass cover

8. The best protection is obtained from **perennial grasses** (*Gramineae*) with the following characteristics:

- fast spread into a dense cover, through creeping, rooting stems (**stolons***) or underground **rhizomes***;
- well adapted to local climate, particularly if seasonally dry;
- easy to propagate vegetatively, for example by transplanting **stolons*** or **rhizomes***.

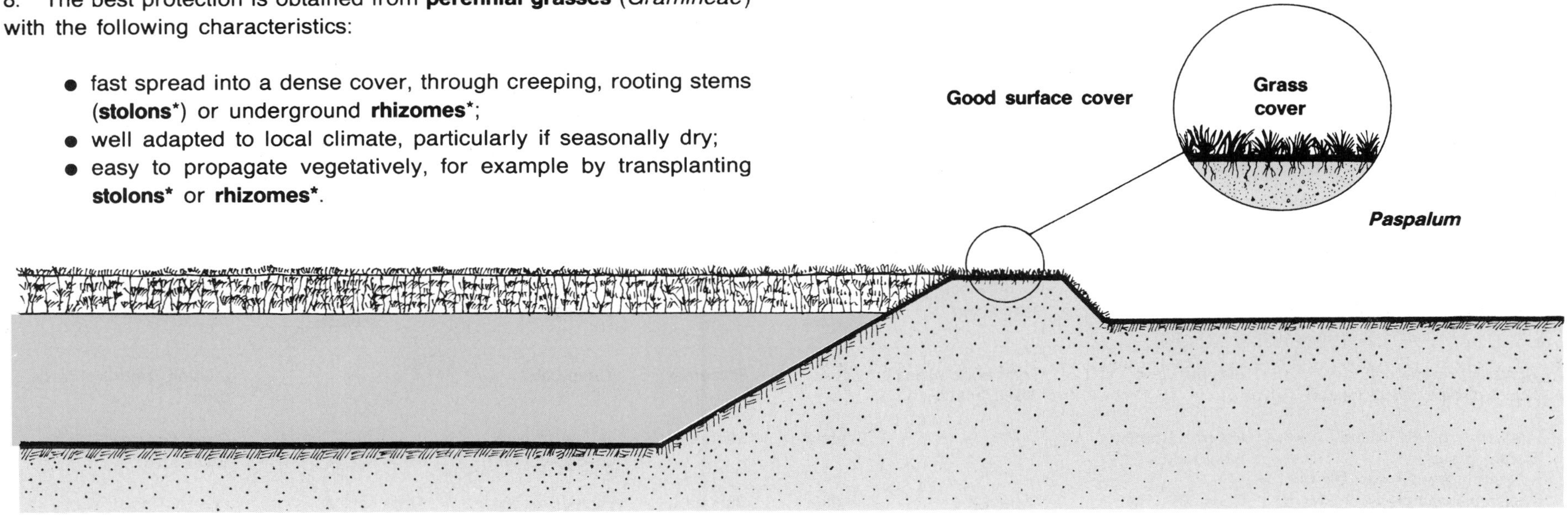

Good surface cover

Grass cover

Paspalum

9. Selected grasses recommended for the formation of a perennial grass cover on pond dikes are listed in **Table 30**.

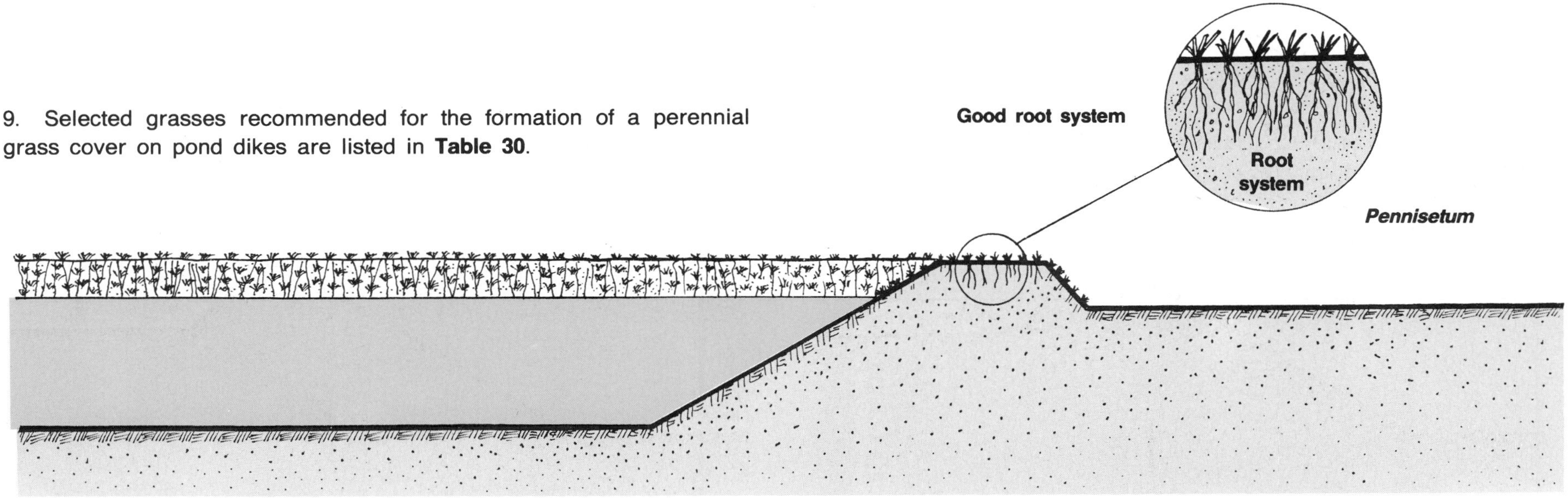

Good root system

Root system

Pennisetum

TABLE 30

Selected perennial grasses for erosion control of dikes

Latin name/common names (* best quality)	Origin/climate	Introduced	Root system	Spread	Climate	Best propagation method	Particular use
Agropyron repens (L.) Couch, twitch, quack grass	Europe, Asia	Temperate areas USA/Canada	—	Rhizomes	Temperate	—	Erosion control of large dikes
Axonopus affinis Chase Carpet grass	Central America West Indies	—	Shallow	Stolons	Humid tropical subtropical	—	Good for moist sandy soils
Axonopus compressus (Swartz) Savannah grass	Coast SE USA West Indies Trop. C + S America	Malaysia Indonesia West Africa	Shallow	Stolons	Humid tropical subtropical	Root sets at 45 × 45 cm	Best on moist, sandy or loamy soils
Brachiaria spp. Tanner grass, parà grass	Tropical Africa Trop. S America	West Indies Venezuela Queensland	—	Stolons	Humid tropical subtropical	Cuttings stems/stolons	Up to 17% protein (dry), best on moist soils
Buchloë dactyloides (Nutt.) Buffalo grass	Great Plains USA	—	—	Stolons	Temperate rain 30 to 60 cm	Pieces of turf	Erosion control on fairly heavy soils
* *Cynodon dactylon* (L.) Bermuda grass, star grass	—	Worldwide	—	Stolons Rhizomes	Tropical subtropical	Cuttings stolons/rhizomes, pieces of turf 90 × 90 cm	Withstands long drought

Selected perennial grasses for erosion control of dikes

Latin name/common names (* best quality)	Origin/climate	Introduced	Root system	Spread	Climate	Best propagation method	Particular use
C. plectostachyus (K. Schum.) Naivasha star grass	East Africa	Zimbabwe	—	Stolons	Tropical	Cuttings, stolons	Good for dry climate (50 to 75 cm rain)
* *Digitaria zwazilandensis* Stent Swaziland finger grass	Southern Africa Swaziland	—	—	Stolons	Trop. - subtropical + summer rains	Cuttings, stolons	—
* *Eremochloa ophiuroides* (Munro) Centipede grass	SE Asia	USA	—	Stolons	Humid tropical and subtropical	Cuttings, stolons	—
* *Paspalum notatum* Flügge Bahia grass	C + S America	USA E + W Africa	Deep	Rhizomes	Trop. - subtropical well distr. rain	Seeds (10 to 22 kg/ha) pieces of turf	Best for sandy soils resists long drought
* *Pennisetum clandestinum* Hochst Kikuyu grass	East Central Africa uplands	Queensland S Africa, Brazil	Deep	Stolons rhizomes	Subtropical + at least 85 cm rain	Pieces of turf 30 × 30 cm	Not adapted for tropics at low altitudes
Stenotaphrum secundatum (Walt.) St Augustine grass, buffalo grass, crab grass	Coasts N, C, S America, West Indies, W + S Africa, SE Asia	—	—	Stolons rhizomes	Humid tropical + subtropical	Stem cuttings 30 × 50 cm	Erosion control in spillways; best on moist soils if drought

610 Pond-bottom drains

1. Pond-bottom drains are ditches that are dug on the bottom of the pond to help the water flow out and to **direct the fish** toward the pond outlet when harvesting.

2. You do not always need bottom drains for your pond, for example in small ponds with a sloping bottom. However, it is better to build bottom drains:

- when the bottom slope is insufficient;
- in large ponds more than 75 m long;
- in barrage ponds with an uneven bottom relief.

Designing the network of drains

3. **Bottom drains** can be designed in different patterns according to the pond bottom topography and shape.

4. If the **bottom topography is fairly even**, it is better to build a **regular network of drains**, for example:

- either **radiating** from the outlet if the pond shape is squarish;
- or in a **fish-bone pattern** if the pond is more elongated.

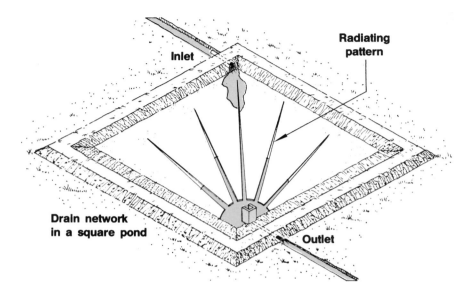

Drain network in a square pond

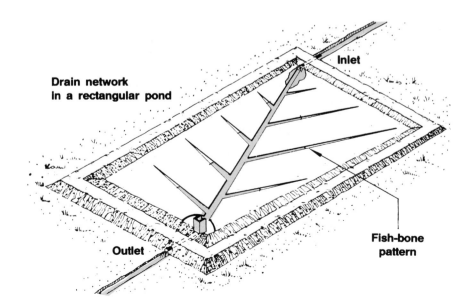

Drain network in a rectangular pond

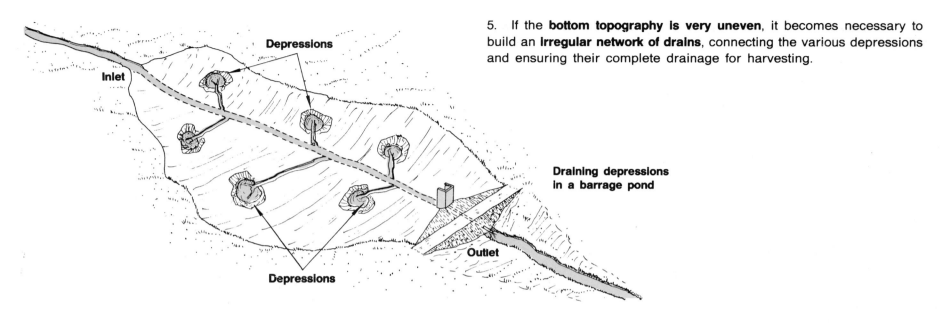

5. If the **bottom topography is very uneven**, it becomes necessary to build an **irregular network of drains**, connecting the various depressions and ensuring their complete drainage for harvesting.

Draining depressions in a barrage pond

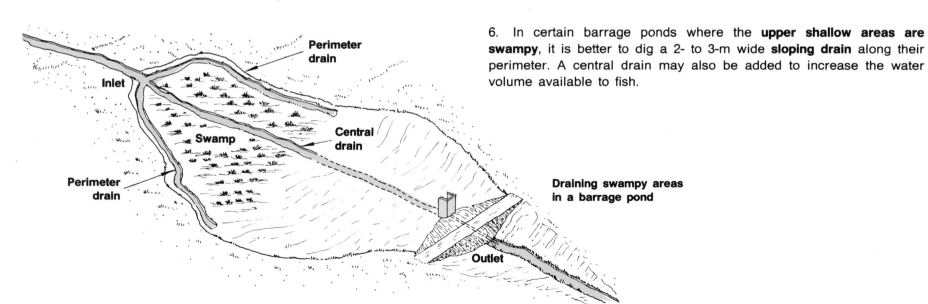

6. In certain barrage ponds where the **upper shallow areas are swampy**, it is better to dig a 2- to 3-m wide **sloping drain** along their perimeter. A central drain may also be added to increase the water volume available to fish.

Draining swampy areas in a barrage pond

7.   In **paddy ponds**, if soil is cut around the inside edge of the pond to
form dikes, the trenches created should be linked in with the outlet drain.

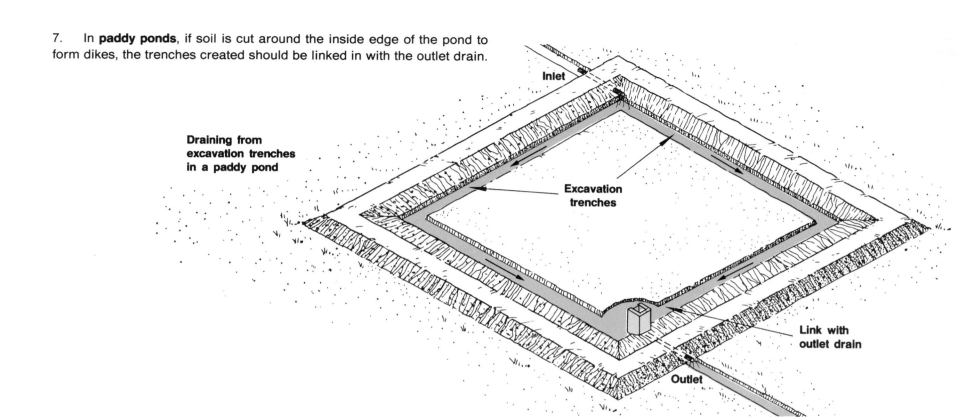

**Draining from
excavation trenches
in a paddy pond**

Inlet

Excavation
trenches

Link with
outlet drain

Outlet

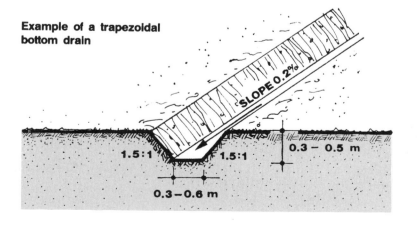

**Example of a trapezoidal
bottom drain**

SLOPE 0.2%

1.5:1   1.5:1   0.3 – 0.5 m

0.3 – 0.6 m

8.   Bottom drains are usually designed as **trapezoidal canals** (see
Section 91, **Pond construction**, **20/2**) with the following characteristics:

- bottom width   0.3 to 0.6 m
- side slopes   1.5:1
- depth   0.3 to 0.5 m
- bottom slope   0.2 percent

9.   The **distance between drains** should vary from 4 to 8 m in small
ponds, to 30 to 50 m in very large ponds. The number of drains should be
kept to the minimum required to completely drain the water, as they will
have to be regularly cleaned by hand.

10. The bottom drains should all communicate with a **harvesting sump** dug in the deepest part of the pond, usually in front of the outlet, where all the fish can be concentrated for their harvest (see **Pond construction, 20/2**).

Bottom drain with harvesting sump

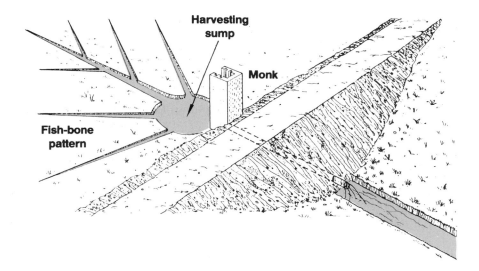

Harvesting sump

Monk

Fish-bone pattern

Note: a **difference in elevation** should be provided:

- between the end of the bottom drain(s) and the bottom of the harvesting sump: at least 20 cm;
- between the bottom of the harvesting sump and the bottom of the outlet structure: at least 10 cm.

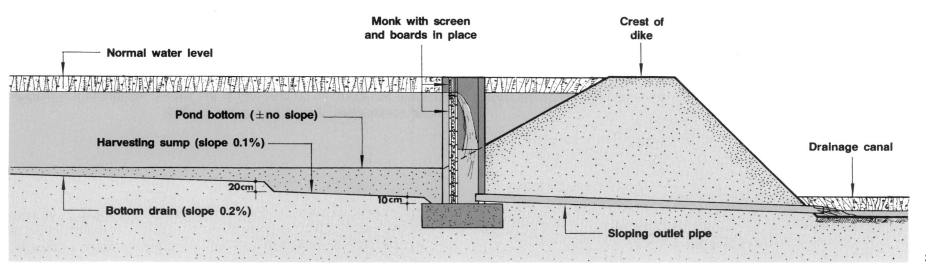

Normal water level

Monk with screen and boards in place

Crest of dike

Pond bottom (±no slope)

Harvesting sump (slope 0.1%)

Drainage canal

20cm

10cm

Bottom drain (slope 0.2%)

Sloping outlet pipe

611 First filling of the pond

1. As soon as possible and before the completion of the pond, it is advisable to put it under water:

- to check that all structures function properly such as the water intake, the canals, the pond inlet and outlet;
- to check that the new dikes are strong and impervious enough;
- to accelerate the stabilization of these dikes.

2. **For maximum security and efficiency**, proceed in the following way.

(a) Fill the pond with water very slowly and up to a maximum depth of 0.40 m at the outlet.

(b) Close the water supply and keep water in the pond for a few days. During this period, check the dikes carefully. Repair crevices and collapsed sections, compacting well.

(c) Drain the water completely and leave the pond dry for a few days. Keep checking the dikes and repair them as necessary.

(d) Fill the pond again very slowly and up to a maximum level about 0.40 m higher than the previous time.

(e) Close the water supply. Check the dikes and repair them as necessary. After a few days, drain the pond completely.

(f) Repeat this process of filling/drying until the water level in the pond reaches the designed maximum level.

(g) Check and repair the dikes as necessary.

3. If there is a **limited water supply**, it will not be possible to follow the above procedure. In such a case, fill the pond very slowly and gradually, closing the water supply at regular intervals and checking on the dikes carefully.

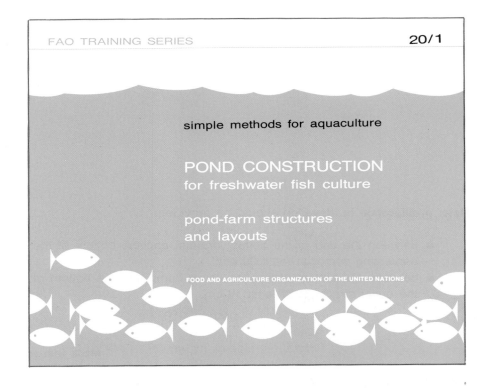

FAO TRAINING SERIES 20/1

simple methods for aquaculture

POND CONSTRUCTION
for freshwater fish culture

pond-farm structures
and layouts

FOOD AND AGRICULTURE ORGANIZATION OF THE UNITED NATIONS

Note: now you are ready to learn more about the various structures which will enable you to better control the supply and transport of water on your fish farm. To do this, turn to the second volume of this manual, **Pond construction,** *FAO Training Series,* **20/2.**

MEASUREMENT UNITS

Length/distance	**m**	= metre
	cm	= centimetre = 0.01 m
	mm	= millimetre = 0.001 m
	in	= inch = 2.54 cm
Area	**m²**	= square metre
	ha	= hectare = 10 000 m²
	cm²	= square centimetre = 0.0001 m²
	mm²	= square millimetre = 0.000001 m²
Volume/capacity	**l**	= litre
	m³	= cubic metre = 1 000 l
Weight/pull	**kg**	= kilogram
	t	= tonne = 1 000 kg

Time	**s**	= second
	min	= minute = 60 s
	h	= hour = 60 min = 3 600 s
Velocity/speed	**m/s**	= metre per second = 3.6 km/h
	km/h	= kilometre per hour = 0.278 m/s
Discharge/flow	**l/s**	= litre per second
	m³/s	= cubic metre per second = 1 000 l/s
	m³/h	= cubic metre per hour = 86.4 l/s
Power	**kW**	= kilowatt = 1.341 HP
	HP	= horsepower = 0.746 kW
	>	greater than
	<	less than

COMMON ABBREVIATIONS

Water discharge

Q	water discharge capacity
D	inside diameter of pipe
H	head or head loss
L	length
TEL	total equivalent length of pipeline
V	water velocity
Vmax	maximum water velocity
M	velocity modulus (**Table 15**)
K	conveyance factor (**Table 15**)

Dike construction

CH	construction height
DH	design height
SA	settlement allowance
SD	slope of dry side of dike
SW	slope of wet side of dike

Water pump

P	power
E	efficiency
HP	horsepower
kW	kilowatt
H	total head
h_d	delivery head
h_s	suction head
h_p	pipe loss head

Ropes and chains

BL	breaking load
SF	safety factor
SWL	safe working load

GLOSSARY OF TECHNICAL TERMS[1]

AGGREGATE
Hard materials such as sand, gravel or stone used for mixing with a cementing material to form mortar or concrete

CENTRE LINE
Longitudinal axis of a dike

CONTOUR LINE
A drawn line which joins points of equal elevation on a plan or a map; it represents a contour as it is found in the field

CUT
Area where it is needed to lower the land to a required elevation by digging soil away

DENSITY
Ratio of the weight of a certain volume of a material to the weight of the same volume of pure water

EROSION
Wear or scouring caused by the abrasive action of moving water or wind

EXPANSION
Increase of the earth volume when disturbed or excavated (synonym: bulking)

FILL
Area where it is needed to raise the land to a required elevation by bringing soil in

FREEBOARD
Upper part of a dike, which should never be under water

GRAVITY
Physical force pulling things toward the centre of the earth; in practical terms, for water, dropping from a higher to a lower elevation

HEAD LOSS
The loss of head, such as that occurring through friction or change of speed, as water is pushed through a pipe or other hydraulic structure

HYDRAULIC GRADIENT
Slope of the water surface within a wet earthen dike (see **saturation line***)

---

[1] This glossary contains definitions of the technical terms marked with an asterisk (*) in the text.

GLOSSARY OF TECHNICAL TERMS, continued

MULCHING | Covering newly planted areas with a protective layer of vegetal material such as straw or leaves

RHIZOME | Thick, horizontal plant stem usually under ground, sending out shoots above and roots below

SATURATION LINE | Upper limit of the wet zone within an earthen dike when partly under water

SETTLEMENT | Downward movement of the soil within a **fill*** such as a dike, because of natural factors (e.g. rain and weight of above soil) and artificial factors (e.g. **compaction** and weight of a vehicle)

SLAB | Flat, usually horizontal, moulded layer of plain or reinforced concrete, usually of uniform thickness

SLAKED LIME | A lime putty obtained by adding water to quicklime

STOLON | Creeping plant stem or runner capable of developing roots and stem to ultimately form a new individual plant

TEMPLATE | A board frame used as a guide in bringing a canal, a dike, etc. to the desired shape

TOE/TOELINE | Base of a dike, away from the dike material; **inside toe**: inside the pond; **outside toe**: outside the pond

FURTHER READING

FAO. 1972. *Manual on the employment of draught animals in agriculture.* Rome, FAO. 249 p.

FAO/UNDP. 1984. Inland aquaculture engineering. *Lectures presented at the ADCP Inter-regional Training Course on Inland Aquaculture Engineering.* Budapest, 6 June - 3 September 1983. ADCP/REP/84/21. Rome, FAO. 591 p.

Intermediate Technology, GTZ/GATE. 1985. *Tools for agriculture. A buyer's guide to appropriate equipment.* London, Intermediate Technology Publications Ltd., 3rd ed. 264 p.

Nichols, Jr., H.L. 1955 *Moving the earth: the workbook of excavation.* Greenwich, Connecticut, North Castle Books.

Société Grenobloise d'Etudes et d'Applications Hydrauliques (SOGREAH). 1975. *Manuel de l'adjoint technique du génie rural: travaux sur un périmètre d'irrigation.* Paris, Ministère de la Coopération. 381 p.

Stern, P. *et al.*, eds. 1983. *Field engineering. An introduction to development work and construction in rural areas.* London, Intermediate Technology Publications Ltd. 251 p.

US Department of Agriculture, Soil Conservation Service. 1982. Ponds. Planning, design and construction. *Agriculture Handbook,* 590. 51 p.

Anonymous. 1974. *Les ouvrages en gabions. Techniques rurales en Afrique.* Paris, Secrétariat d'Etat aux Affaires Etrangères. 122 p.

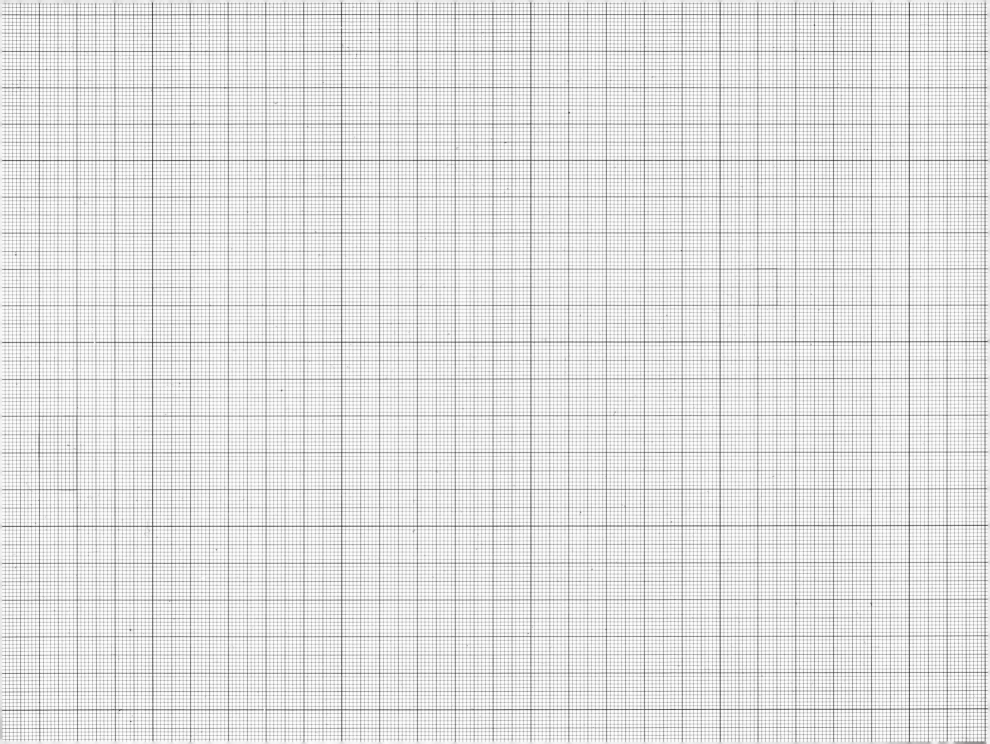

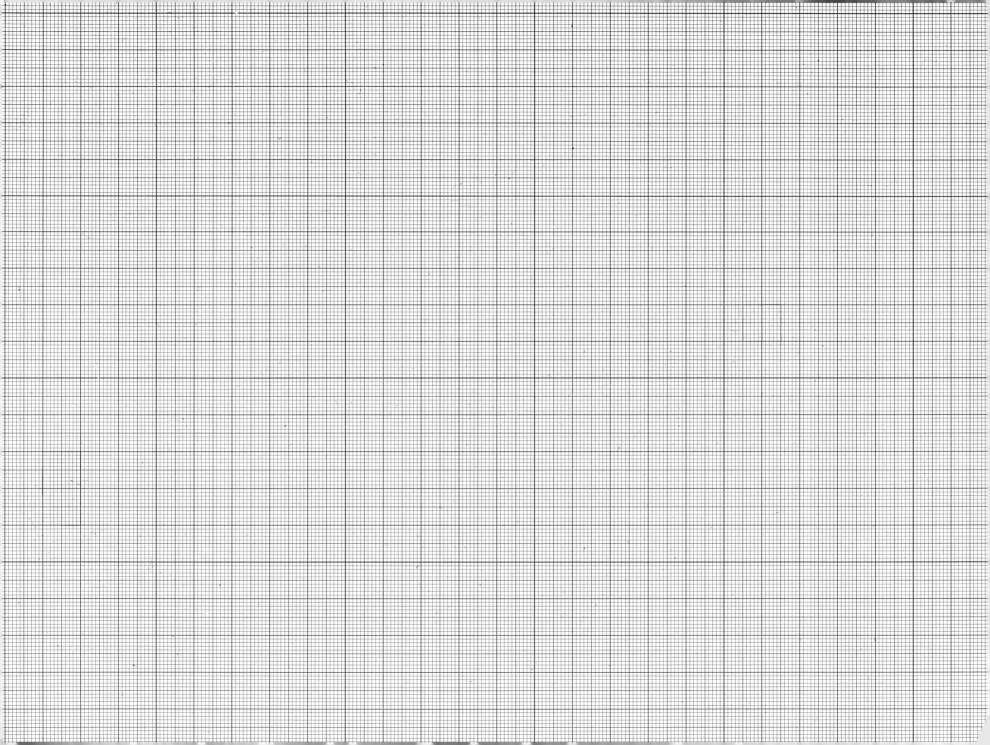

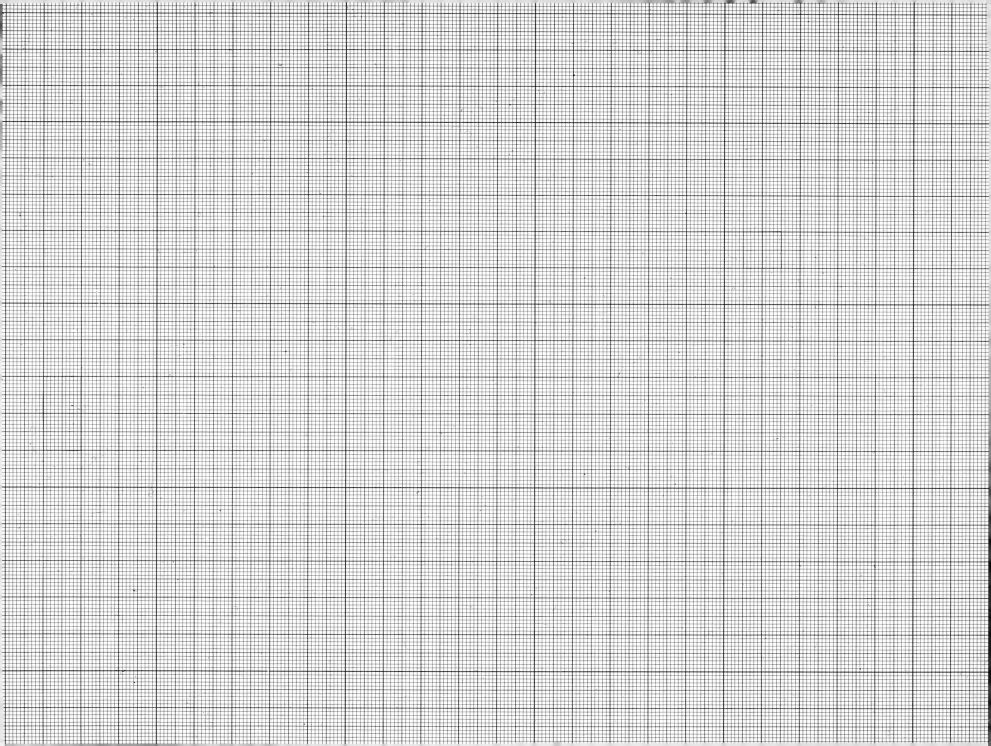

NOTES

NOTES

NOTES

NOTES

NOTES

WHERE TO PURCHASE FAO PUBLICATIONS LOCALLY
POINTS DE VENTE DES PUBLICATIONS DE LA FAO
PUNTOS DE VENTA DE PUBLICACIONES DE LA FAO

• ANGOLA
Empresa Nacional do Disco e de
Publicações, ENDIPU-U.E.E.
Rua Cirilo da Conceição Silva, Nº 7
C.P. Nº 1314-C
Luanda

• ARGENTINA
Librería Agropecuaria
Pasteur 743
1028 Buenos Aires
Oficina del Libro Internacional
Alberti 40
1082 Buenos Aires

• AUSTRALIA
Hunter Publications
P.O. Box 404
Abbotsford, Vic. 3067

• AUSTRIA
Gerold Buch & Co.
Weihburggasse 26
1010 Vienna

• BANGLADESH
Association of Development
Agencies in Bangladesh
House Nº 1/3, Block F, Lalmatia
Dhaka 1207

• BELGIQUE
M.J. De Lannoy
202, avenue du Roi
1060 Bruxelles
CCP 000-0808993-13

• BOLIVIA
Los Amigos del Libro
Perú 3712, Casilla 450
Cochabamba
Mercado 1315, La Paz

• BOTSWANA
Botsalo Books (Pty) Ltd
P.O. Box 1532
Gaborone

• BRAZIL
Fundação Getúlio Vargas
Praia do Botafogo 190, C.P. 9052
Rio de Janeiro

• CANADA
(See North America)

• CHILE
Librería - Oficina Regional FAO
Avda. Santa María 6700
Casilla 10095, Santiago
Tel.   218 53 23
Fax   218 25 47
Universitaria Textilibros Ltda.
Avda. L. Bernardo O'Higgins 1050
Santiago

• CHINA
China National Publications Import &
Export Corporation
P.O. Box 88
100704 Beijing

• COLOMBIA
Banco Ganadero
Revista Carta Ganadera
Carrera 9ª Nº 72-21, Piso 5
Bogotá D.E.
Tel. 217 0100

• CONGO
Office national des librairies populaires
B.P. 577
Brazzaville

• COSTA RICA
Librería, Imprenta y Litografía Lehmann
S.A.
Apartado 10011
San José

• CÔTE D'IVOIRE
CEDA - Centre d'édition
et de diffusion africaines
04 B.P. 541
Abidjan 04

• CUBA
Ediciones Cubanas, Empresa de
Comercio Exterior de
Publicaciones
Obispo 461, Apartado 605
La Habana

• CYPRUS
MAM
P.O. Box 1722
Nicosia

• CZECH REPUBLIC
Artia
Ve Smeckach 30, P.O. Box 790
11127 Praha 1
Artia Pegas Press Ltd
Import of Periodicals
Palác Metro, P.O. Box 825
Národni 25, 11121 Praha 1

• DENMARK
Munksgaard, Book and Subscription
Service
P.O. Box 2148
DK 1016 Copenhagen K.
Tel. 4533128570
Fax 4533129387

• ECUADOR
Libri Mundi, Librería Internacional
Juan León Mera 851
Apartado Postal 3029
Quito

• EGYPT
The Middle East Observer
41 Sherif Street
Cairo

• ESPAÑA
Mundi Prensa Libros S.A.
Castelló 37
28001 Madrid
Tel. 431 3399
Fax 575 3998
Librería Agrícola
Fernando VI 2
28004 Madrid
Librería Internacional AEDOS
Consejo de Ciento 391
08009 Barcelona
Tel. 301 8615
Fax 317 0141
Librería de la Generalitat de Catalunya
Rambla dels Estudis, 118
(Palau Moja)
08002 Barcelona
Tel. (93) 302 6462
Fax (93) 302 1299

• FINLAND
Akateeminen Kirjakauppa
P.O. Box 218
SF-00381 Helsinki

• FRANCE
La Maison Rustique
Flammarion 4
26, rue Jacob
75006 Paris
Lavoisier
14, rue de Provigny
94236 Cachan Cedex
Librairie de l'Unesco
7, place de Fontenoy
75700 Paris
Editions A. Pedone
13, rue Soufflot
75005 Paris
Librairie du Commerce International
24, boulevard de l'Hôpital
75005 Paris

• GERMANY
Alexander Horn Internationale
Buchhandlung
Kirchgasse 22, Postfach 3340
D-65185 Wiesbaden
Uno-Verlag
Poppelsdorfer Allee 55
D-53115 Bonn 1
S. Toeche-Mittler GmbH
Versandbuchhandlung
Hindenburgstrasse 33
D-64295 Darmstadt

• GHANA
SEDCO Publishing Ltd
Sedco House, Tabon Street
Off Ring Road Central, North Ridge
P.O. Box 2051
Accra

• GREECE
G.C. Eleftheroudakis S.A.
4 Nikis Street
10563 Athens
John Mihalopoulos & Son S.A.
75 Hermou Street, P.O. Box 10073
75110 Thessaloniki

• GUYANA
Guyana National Trading
Corporation Ltd
45-47 Water Street, P.O. Box 308
Georgetown

• HAÏTI
Librairie "A la Caravelle"
26, rue Bonne Foi, B.P. 111
Port-au-Prince

• HONDURAS
Escuela Agrícola Panamericana, Librería
RTAC
Zamorano, Apartado 93
Tegucigalpa
Oficina de la Escuela Agrícola
Panamericana en Tegucigalpa
Blvd. Morazán, Apts. Glapson - Apartado
93
Tegucigalpa

• HONG KONG
Swindon Book Co.
13-15 Lock Road
Kowloon

• HUNGARY
Librotrade Kft.
P.O. Box 126
H-1656 Budapest

• INDIA
EWP Affiliated East-West Press PVT, Ltd
G-I/16, Ansari Road
Darya Ganj
New Delhi 110 002
Oxford Book and Stationery Co.
Scindia House, New Delhi 110 001;
17 Park Street, Calcutta 700 016
Oxford Subscription Agency
Institute for Development Education
1 Anasuya Ave., Kilpauk
Madras 600 010

• IRAN
The FAO Bureau, International and
Regional Specialized Organizations
Affairs
Ministry of Agriculture of the Islamic
Republic of Iran
Keshavarz Bld., M.O.A., 17th floor Teheran

• IRELAND
Publications Section, Stationery Office
4-5 Harcourt Road
Dublin 2

• ISRAEL
R.O.Y. International
P.O. Box 13056
Tel Aviv 61130

• ITALY
FAO (See last column)
Libreria Scientifica Dott. Lucio de Biasio
"Aeiou"
Via Coronelli 6
20146 Milano
Libreria Concessionaria Sansoni S.p.A.
"Licosa"
Via Duca di Calabria 1/1
50125 Firenze
Libreria Internazionale Rizzoli
Galleria Colonna, Largo Chigi
00187 Roma
FAO Bookshop
Viale delle Terme di Caracalla
00100 Roma

• JAPAN
Far Eastern Booksellers
(Kyokuto Shoten Ltd)
12 Kanda-Jimbocho 2 chome
Chiyoda-ku - P.O. Box 72
Tokyo 101-91
Maruzen Company Ltd
P.O. Box 5050
Tokyo International 100-31

• KENYA
Text Book Centre Ltd
Kijabe Street, P.O. Box 47540
Nairobi

• KUWAIT
The Kuwait Bookshops Co. Ltd
P.O. Box 2942
Safat

• LUXEMBOURG
M.J. De Lannoy
202, avenue du Roi
1060 Bruxelles (Belgique)

• MALAYSIA
Electronic products only:
Southbond
Sendirian Berhad Publishers
9 College Square
10250 Penang

• MALI
Librairie Traore
Rue Soundiata Keita X· 115
B.P. 3243
Bamako

• MAROC
Librairie "Aux Belles Images"
281, avenue Mohammed V
Rabat

• MEXICO
Librería, Universidad Autónoma de
Chapingo
56230 Chapingo
Libros y Editoriales S.A.
Av. Progreso Nº 202-1º Piso A
Apdo Postal 18922 Col. Escandón
11800 México D.F.

• NETHERLANDS
Roodveldt Import B.V.
Brouwersgracht 288
1013 HG Amsterdam

• NEW ZEALAND
Legislation Services
P.O. Box 12418
Thorndon, Wellington

• NICARAGUA
Librería Universitaria,
Universidad Centroamericana
Apartado 69
Managua

• NIGERIA
University Bookshop (Nigeria) Ltd
University of Ibadan
Ibadan

• NORTH AMERICA
Publications:
UNIPUB
4611/F, Assembly Drive
Lanham MD 20706-4391, USA
Toll-free   800 233-
0504 (Canada)
                        800 274-
4888 (USA)
Fax   301-459-
0056
Periodicals:
Ebsco Subscription Services
P.O. Box 1431
Birmingham AL 35201-1431, USA
Tel.   (205)   991-6600
Telex   78-2661
Fax   (205)   991-1449
The Faxon Company Inc.
15 Southwest Park
Westwood MA 02090, USA
Tel.   6117-329-3350
Telex 95-1980
Cable FW Faxon Wood
Electronic products only:
Winrock International Institute for
Agricultural Development
1611 North Kent Street
Suite 600 Service
Arlington VA 22209

• NORWAY
Narvesen Info Center
Bertrand Narvesens vei 2
P.O. Box 6125, Etterstad
0602 Oslo 6

• PAKISTAN
Mirza Book Agency
65 Shahrah-e-Quaid-e-Azam
P.O. Box 729, Lahore 3
Sasi Book Store
Zaibunnisa Street
Karachi

• PERU
Librería Distribuidora "Santa Rosa"
Jirón Apurímac 375, Casilla 4937
Lima 1

• PHILIPPINES
International Book Center (Phils)
Room 1703, Cityland 10
Condominium Cor. Ayala Avenue &
H.V. dela Costa Extension
Makati, Metro Manila

• POLAND
Ars Polona
Krakowskie Przedmiescie 7
00-950 Warsaw

• PORTUGAL
Livraria Portugal,
Dias e Andrade Ltda.
Rua do Carmo 70-74, Apartado 2681
1117 Lisboa Codex

• SINGAPORE
Select Books Pte Ltd
03-15 Tanglin Shopping Centre
19 Tanglin Road
Singapore 1024

• SLOVENIA
Cankarjeva Zalozba
P.O. Box 201-IV
61001 Ljubljana

• SOMALIA
"Samater's"
P.O. Box 936
Mogadishu

• SRI LANKA
M.D. Gunasena & Co. Ltd
217 Olcott Mawatha, P.O. Box 246
Colombo 11

• SUISSE
Librairie Payot S.A.
107 Freiestrasse, 4000 Basel 10
6, rue Grenus, 1200 Genève
Case Postale 3212, 1002 Lausanne
Buchhandlung und Antiquariat
Heinimann & Co.
Kirchgasse 17
8001 Zurich
UN Bookshop
Palais des Nations
CH-1211 Genève 1
Van Diermen Editions Techniques
ADECO
Case Postale 465
CH-1211 Genève 19

• SURINAME
Vaco n.v. in Suriname
Domineestraat 26, P.O. Box 1841
Paramaribo

• SWEDEN
Books and documents:
C.E. Fritzes
P.O. Box 16356
103 27 Stockholm
Subscriptions:
Vennergren-Williams AB
P.O. Box 30004
104 25 Stockholm

• THAILAND
Suksapan Panit
Mansion 9, Rajdamnern Avenue
Bangkok

• TOGO
Librairie du Bon Pasteur
B.P. 1164
Lomé

• TUNISIE
Société tunisienne de diffusion
5, avenue de Carthage
Tunis

• TURKEY
Kultur Yayiniari is - Turk Ltd Sti.
Ataturk Bulvari No. 191, Kat. 21
Ankara
Bookshops in Istambul and Izmir

• UNITED KINGDOM
HMSO Publications Centre
51 Nine Elms Lane
London SW8 5DR
Tel.   (071) 873 9090 (orders)
       (071) 873 0011 (inquiries)
Fax   (071) 873 8463
and through HMSO Bookshops
Electronic products only:
Microinfo Ltd
P.O. Box 3, Omega Road, Alton
Hampshire GU342PG
Tel.   (0420) 86848
Fax   (0420) 89889

• URUGUAY
Librería Agropecuaria S.R.L.
Buenos Aires 335
Casilla 1755
Montevideo C.P. 11000

• USA
(See North America)

• VENEZUELA
Tecni-Ciencia Libros S.A.
Torre Phelps-Mezzanina
Plaza Venezuela
Caracas
Tel. 782 8697-781 9945-781 9954
Tamanaco Libros Técnicos S.R.L.
Centro Comercial Ciudad Tamanaco, Nivel
C-2
Caracas
Tel. 261 3344-261 3335-959 0016
Tecni-Ciencia Libros S.A.
Centro Comercial, Shopping Center
Av. Andrés Eloy, Urb. El Prebo
Valencia, Edo. Carabobo
Tel. 222 724
Fudeco, Librería
Avenida Libertador-Este
Ed. Fudeco, Apartado 254
Barquisimeto C.P. 3002, Ed. Lara
Tel.   (051) 538 022
Fax   (051) 544 394
Télex (051) 513 14 FUDEC VC

• YUGOSLAVIA
Jugoslovenska Knjiga, Trg.
Republike 5/8, P.O. Box 36
11001 Belgrade
Prosveta
Terazije 16/1
Belgrade

• ZIMBABWE
Grassroots Books
100 Jason Moyo Avenue
P.O. Box A 267, Avondale
Harare;
61a Fort Street
Bulawayo

Other countries / Autres pays / Otros paises
Distribution and Sales Section, FAO
Viale delle Terme di Caracalla
00100 Rome, Italy
Tel.   (39-6) 52251
Fax   52253152
Telex 625852/625853/610181 FAO I

1/8/94

The following is a list manuals on aquaculture published in the FAO TRAINING SERIES:

"Simple methods for acquaculture" series:

Volume 4 — **Water for freshwater fish culture**
                        1981. 111 pp. ISBN 92-5-101112
Volume 6 — **Soil and freshwater fish culture**
                        1986. 174 pp. ISBN 92-5-101355-1
Volume 16/1 — **Topography for freshwater fish culture: topographical tools**
                        1988. 328 pp. ISBN 92-5-102590-8
Volume 16/2 — **Topography for freshwater fish culture: topographical surveys**
                        1989. 266 pp. ISBN 92-5-102591-6
Volume 20/1 — **Pond construction for freshwater fish culture: building earthen ponds**
                        1995. 355 pp. ISBN 92-5-102645-9
Volume 20/2 — **Pond construction for freshwater fish culture: pond-farm structures and layouts**
                        1992. 214 pp. ISBN 92-5-102872-9

In preparation:

Volume 21/1 — **Management for freshwater fish culture: ponds and water practices**
Volume 21/2 — **Management for freshwater fish culture: farms and fish stocks**

Other manuals on acquaculture in the FAO TRAINING SERIES:

Volume 8 — **Common carp 1: mass production of eggs and early fry**
                        1985. 87 pp. ISBN 92-5-102301-8
Volume 9 — **Common carp 2: mass production of advanced fry and fingerlings in ponds**
                        1985. 85 pp. ISBN 92-5-102302-6
Volume 19 — **Simple economics and bookkeeping for fish farmers**
                        1992. 96 pp. ISBN 92-5-103002-2